ISBN: 9798391026082

Website: JadyAlvarez.com

Email: jdalearning@gmail.com

Youtube Educational Videos: https://www.youtube.com/c/JadyAlvarez

Instagram: JadyAHomeschool

Numbers

0 1 2 3 4

5 6 7 8 9

10 11 12 13

14 15 16 17

18 19 20

Zero

Make the number Zero with your hand.
Trace the number Zero.
Write the number Zero.

One

Make the number One with your hand.
Trace the number One.
Write the number One.

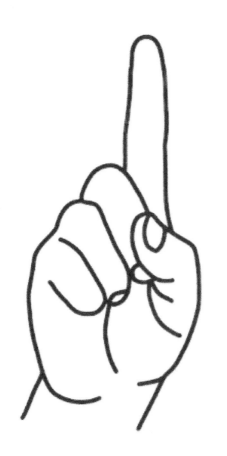

Two

Make the number Two with your hand.
Trace the number Two.
Write the number Two.

Three

Make the number Three with your hand.
Trace the number Three.
Write the number Three.

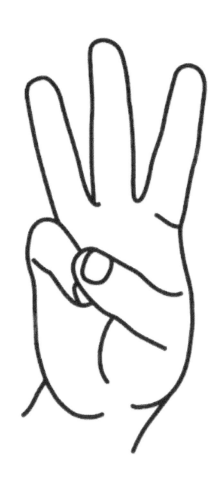

Four

Make the number Four with your hand.
Trace the number Four.
Write the number Four.

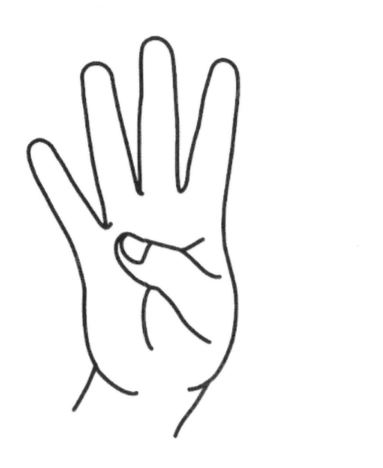

Five

Make the number Five with your hand.
Trace the number Five.
Write the number Five.

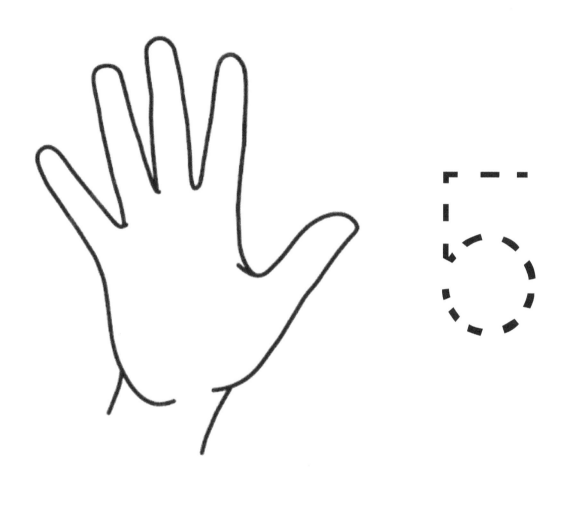

Six

Make the number Six with your hand.
Trace the number Six.
Write the number Six.

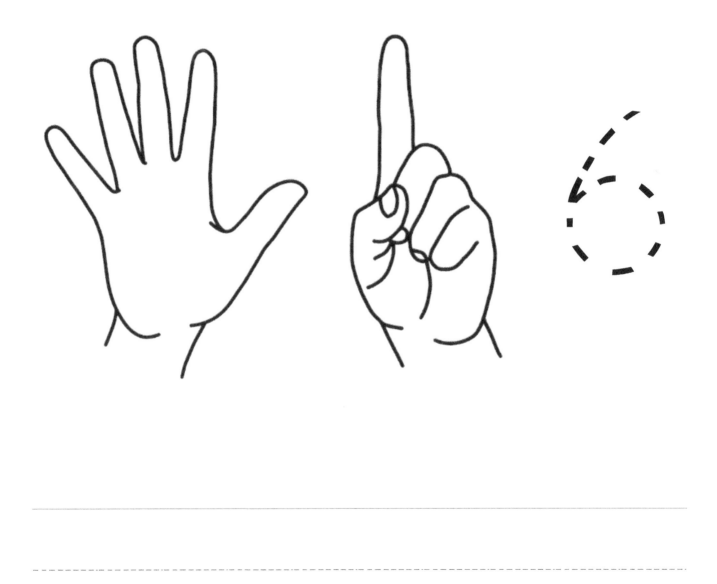

Seven

Make the number Seven with your hand.
Trace the number Seven.
Write the number Seven.

Eight

Make the number Eight with your hand.
Trace the number Eight.
Write the number Eight.

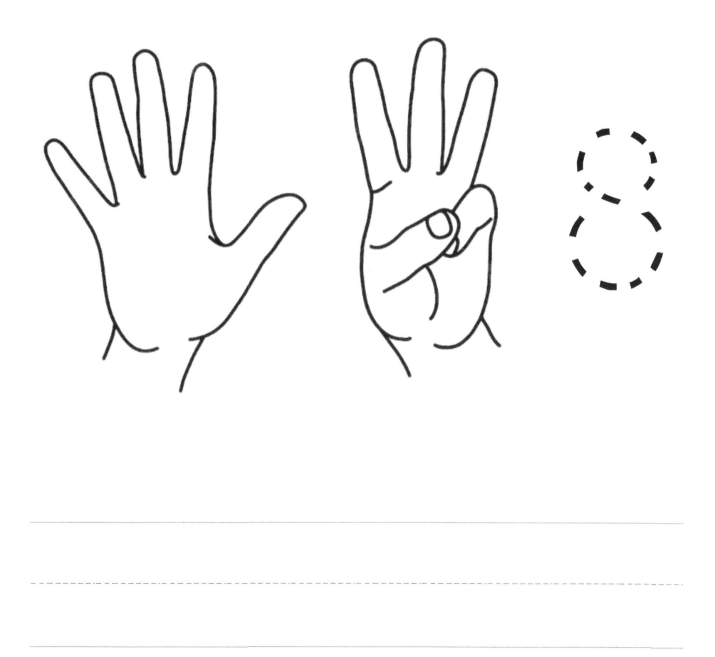

Nine

Make the number Nine with your hand.
Trace the number Nine.
Write the number Nine.

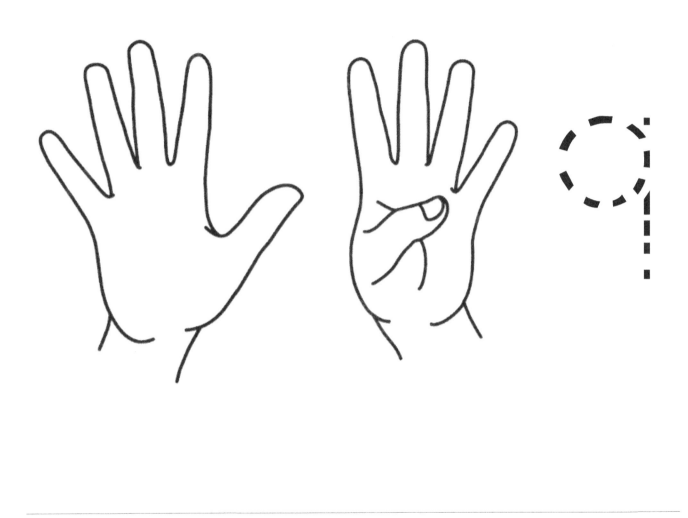

Ten

Make the number Ten with your hand.
Trace the number Ten.
Write the number Ten.

Matching

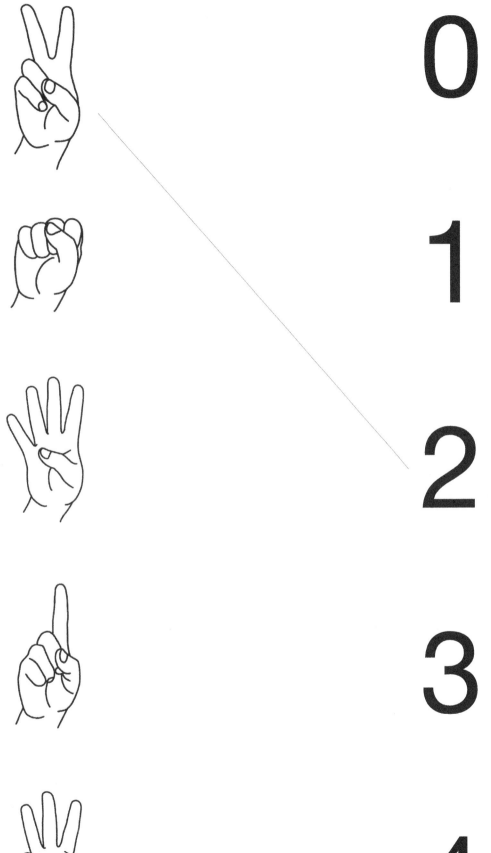

0

1

2

3

4

Matching

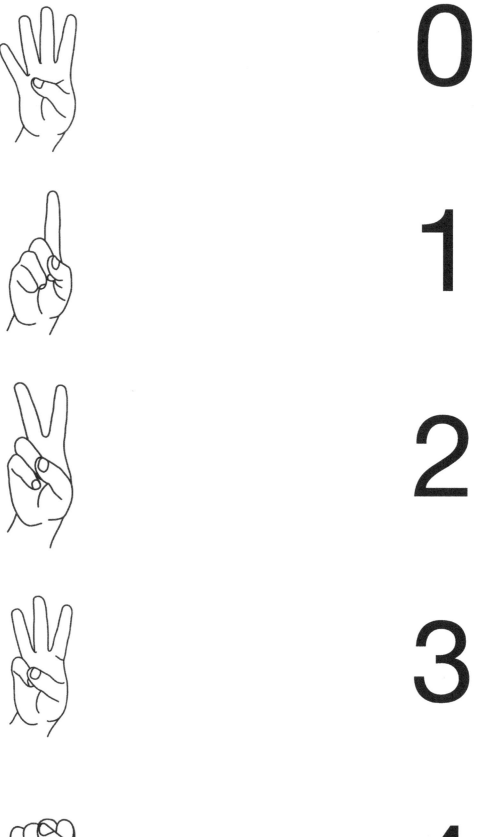

0

1

2

3

4

Matching

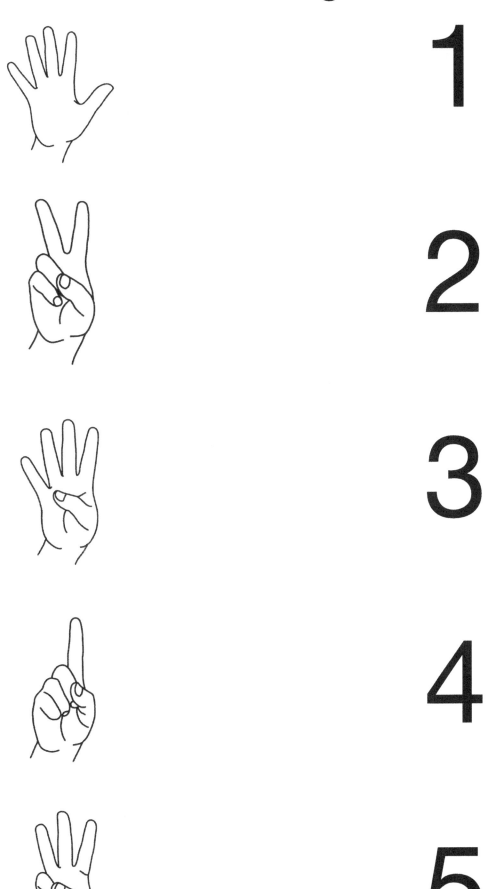

1

2

3

4

5

Matching

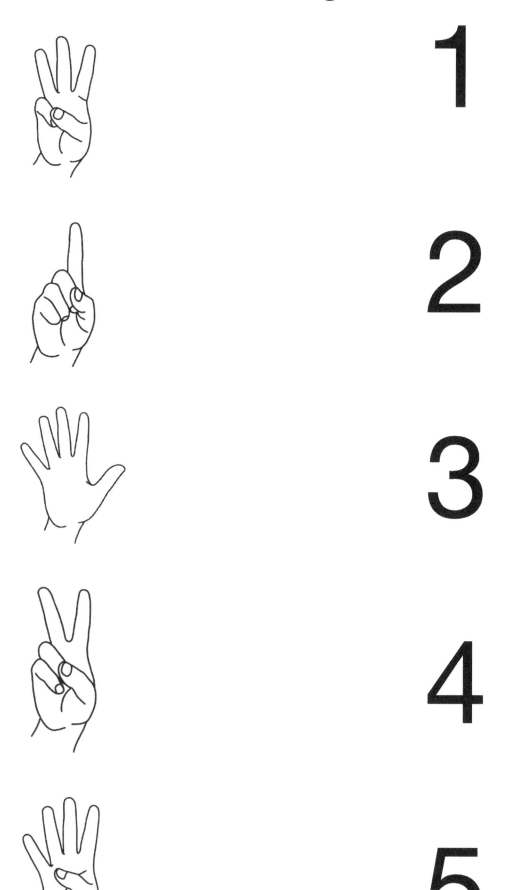

1

2

3

4

5

Matching

6

7

8

9

10

Matching

6

7

8

9

10

Write the Number

_____

_____

_____

_____

_____

_____

Write the Number

_____

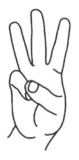

_____

Write the Number

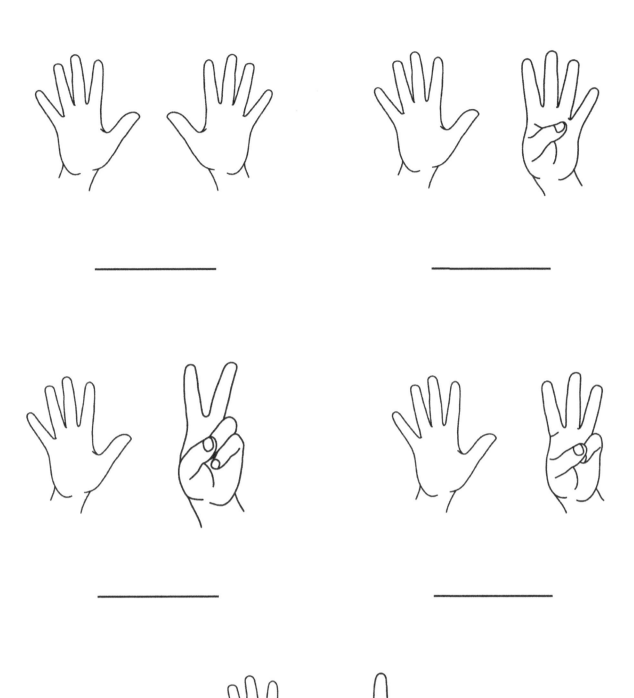

_____   _____

_____   _____

_____

Write the Number

_____

_____

_____

_____

_____

Color the Fingers

Color the correct number of fingers.

2

5

9

3

Color the Fingers

Color the correct number of fingers.

0

1

4

6

Color the Fingers

Color the correct number of fingers.

10

7

3

8

Adding with Fingers Example

Work with the student to learn to add with fingers. Show the child the finger problems and have them tell you the answer.

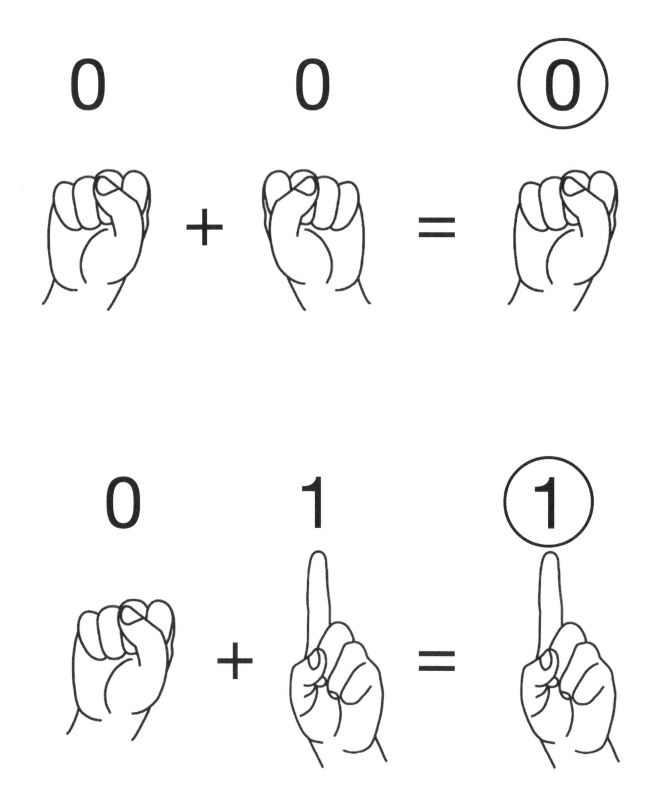

Adding with Fingers Example

Work with the student to learn to add with fingers. Show the child the finger problems and have them tell you the answer.

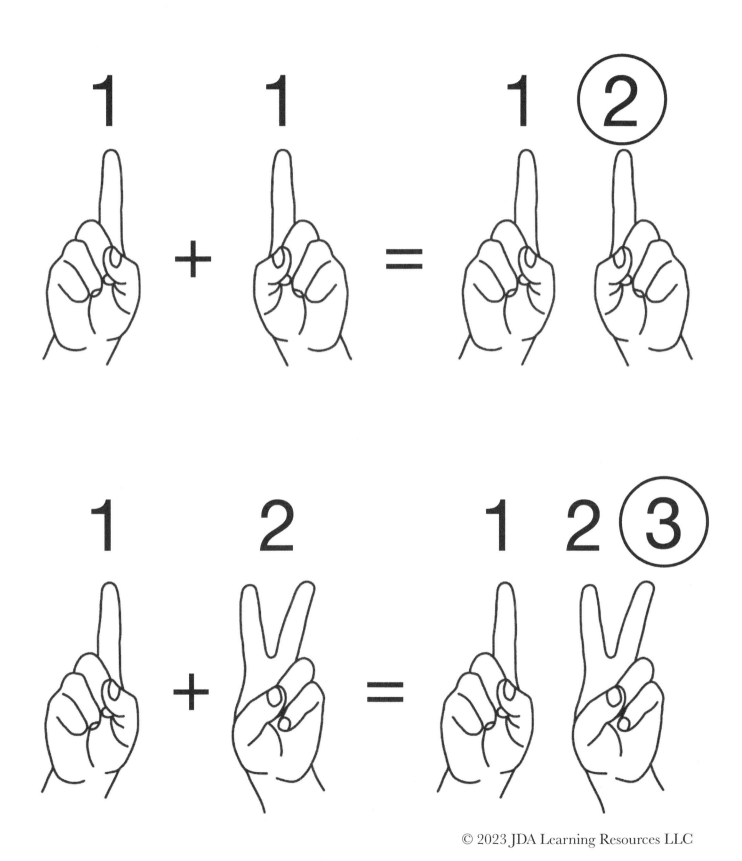

Adding with Fingers Example

Work with the student to learn to add with fingers. Show the child the finger problems and have them tell you the answer.

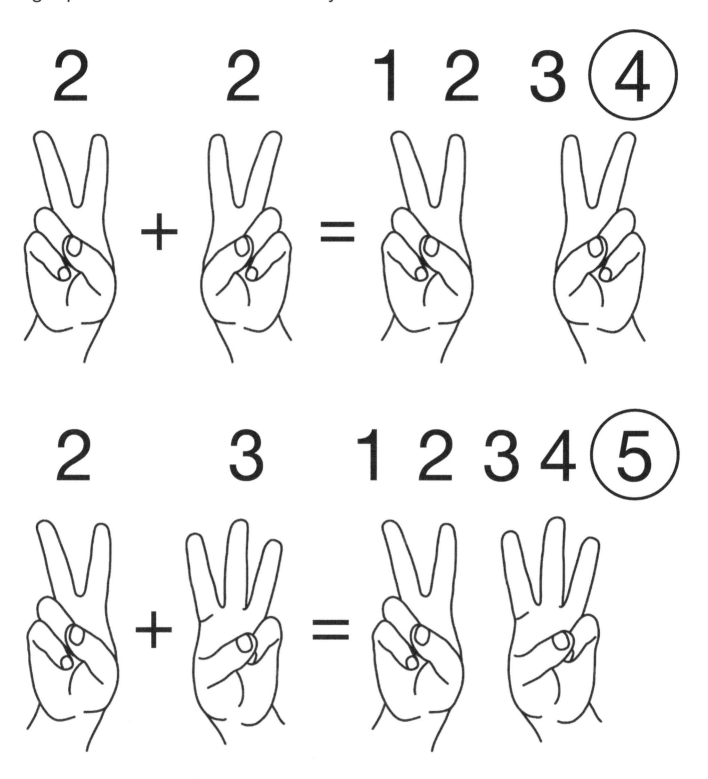

Continue to give the child different addition problems with your hand, and have them give you the answer until they have mastered this concept.

Adding with Fingers

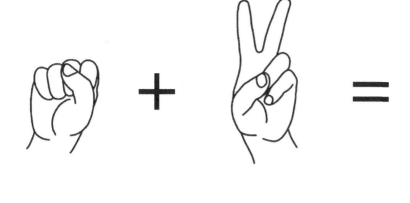

_____ _____   _____

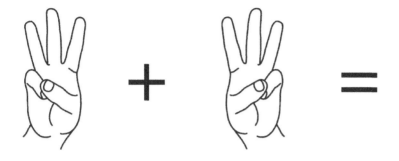

_____ _____   _____

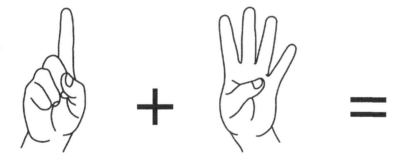

_____ _____   _____

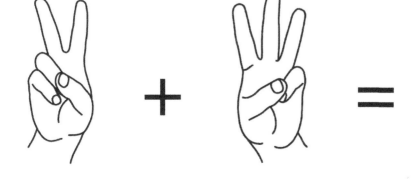

_____ _____   _____

Adding with Fingers

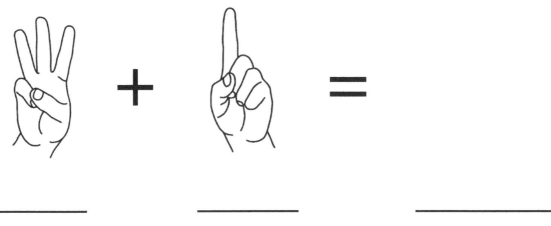

_____   _____   _____

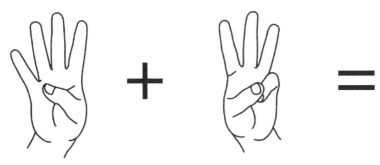

_____   _____   _____

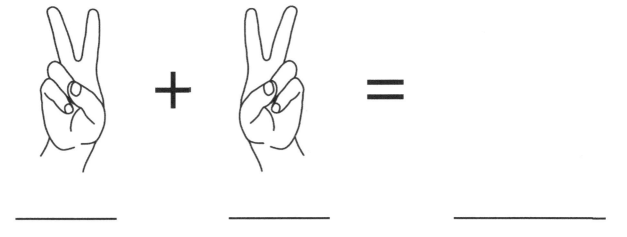

_____   _____   _____

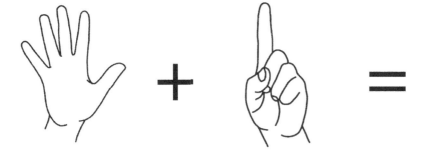

_____   _____   _____

Adding with Fingers

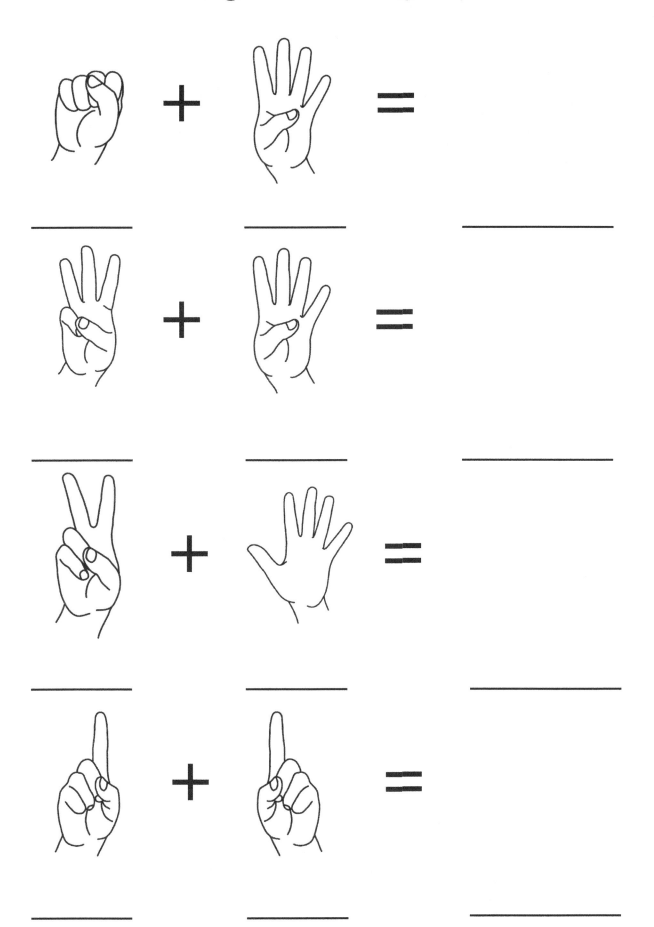

Adding with Fingers

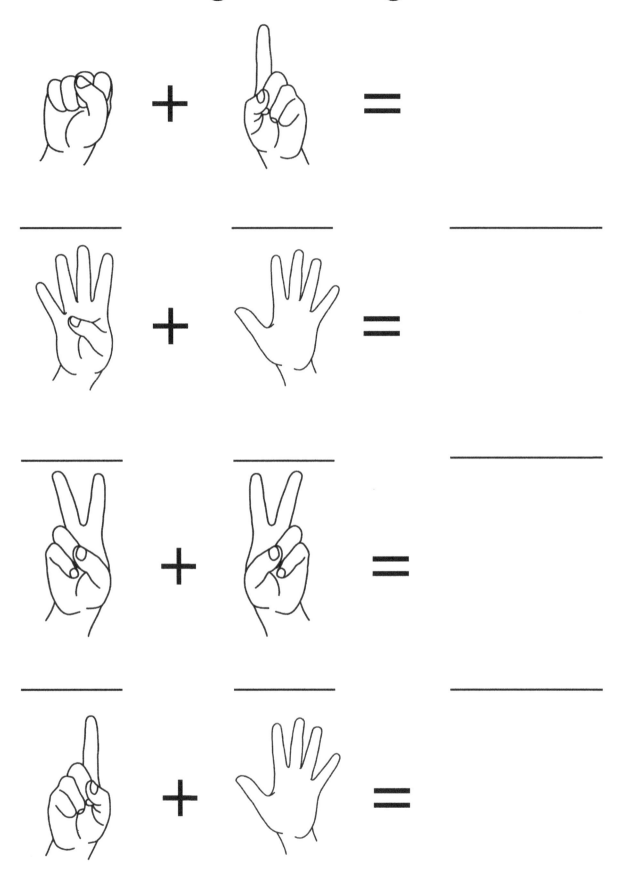

Adding with Fingers

Adding with Fingers

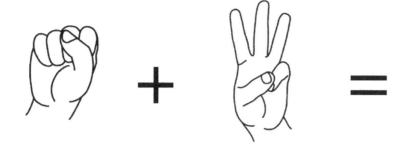

_____   _____   _____

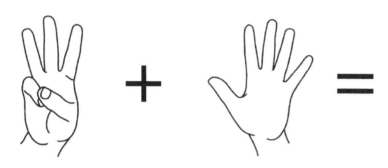

_____   _____   _____

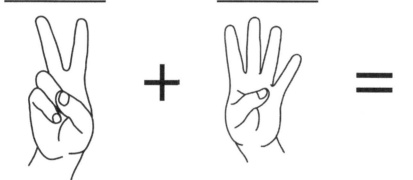

_____   _____   _____

_____   _____   _____

Adding with Fingers

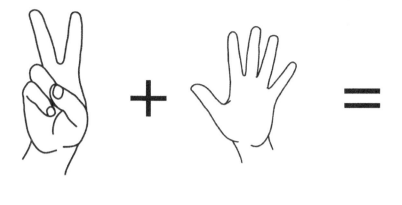

_____   _____   _____

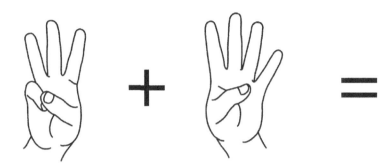

_____   _____   _____

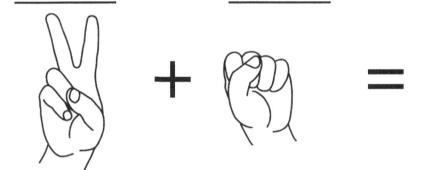

_____   _____   _____

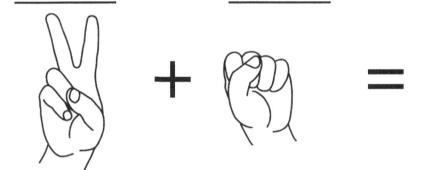

_____   _____   _____

Adding with Fingers

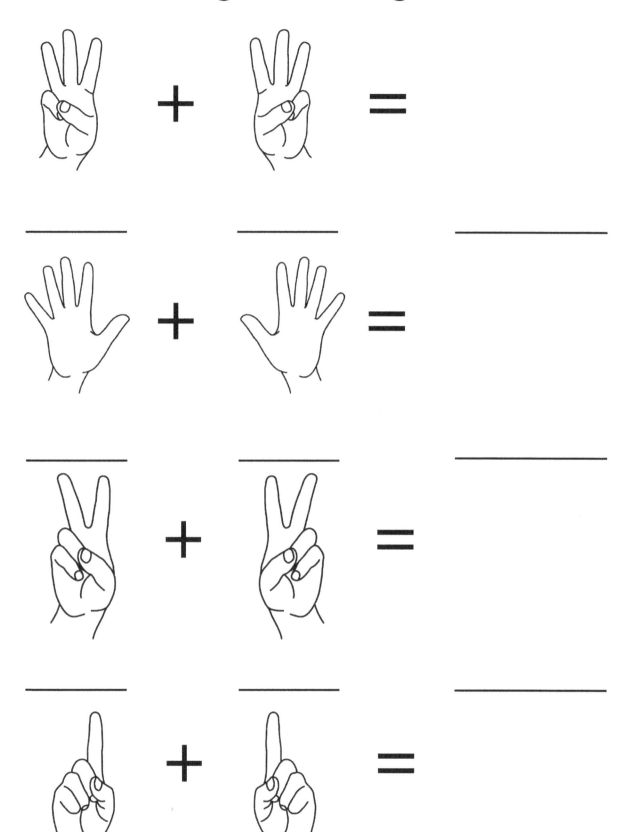

Adding with Fingers

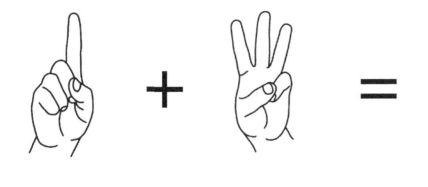

_____   _____   _____

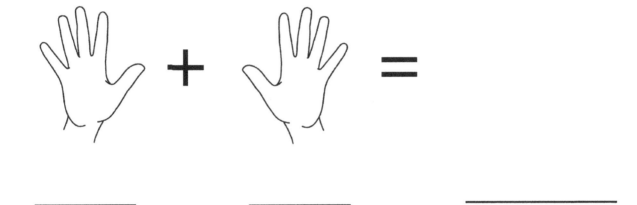

_____   _____   _____

_____   _____   _____

_____   _____   _____

Adding with Fingers

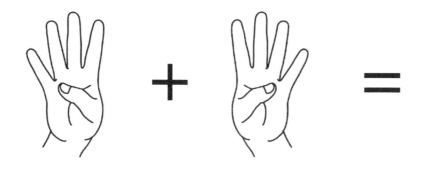

_____   _____   _____

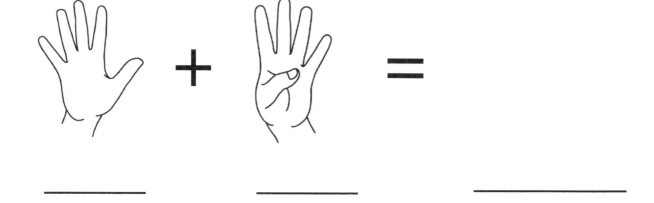

_____   _____   _____

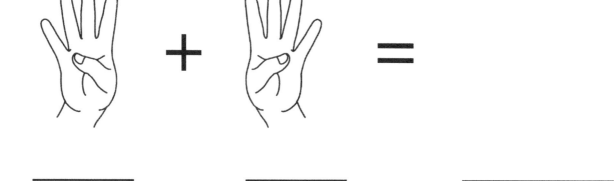

_____   _____   _____

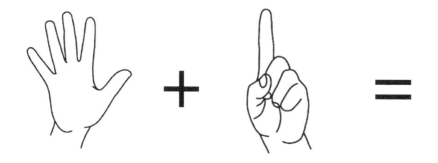

_____   _____   _____

Color and Add

Color the fingers to add.

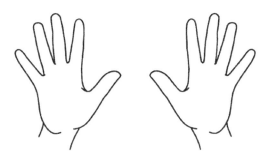 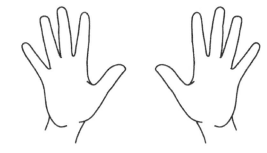

3 + 1 = _____   4 + 2 = _____

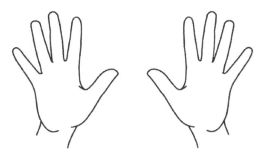

1 + 5 = _____   2 + 2 = _____

 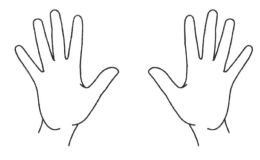

5 + 4 = _____   0 + 3 = _____

Color and Add

Color the fingers to add.

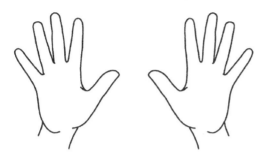 

5 + 3 = _____   2 + 5 = _____

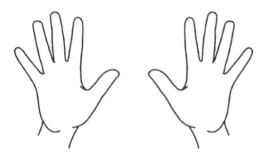

3 + 3 = _____   1 + 0 = _____

 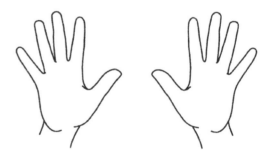

4 + 4 = _____   2 + 1 = _____

Color and Add

Color the fingers to add.

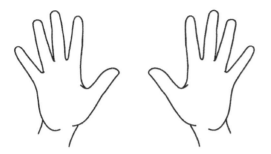

$3 + 0 = $ _____

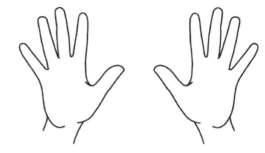

$5 + 4 = $ _____

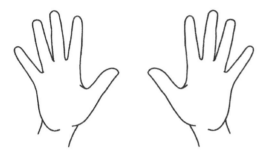

$2 + 3 = $ _____

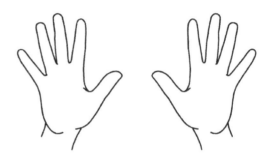

$4 + 5 = $ _____

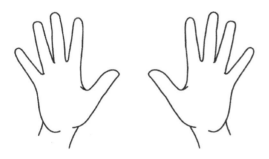

$4 + 1 = $ _____

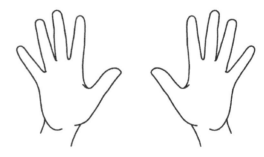

$5 + 0 = $ _____

Color and Add

Color the fingers to add.

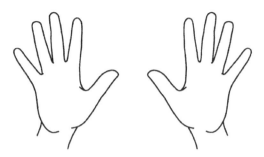

3 + 4 = _____

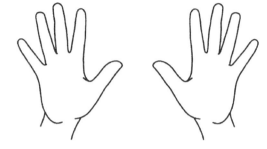

2 + 4 = _____

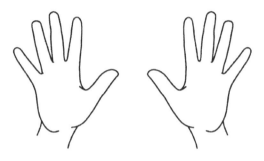

4 + 0 = _____

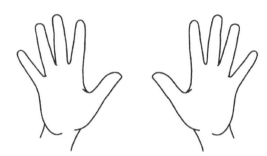

3 + 1 = _____

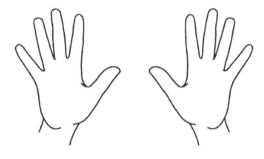

5 + 5 = _____

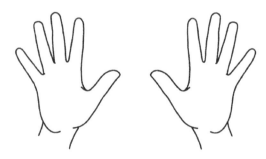

1 + 1 = _____

Addition Problems

0 + 0 = _____   1 + 3 = _____

1 + 5 = _____   0 + 4 = _____

3 + 4 = _____   2 + 3 = _____

5 + 5 = _____   0 + 2 = _____

2 + 5 = _____   1 + 1 = _____

4 + 4 = _____   5 + 5 = _____

Addition Problems

2 + 2 = _____   5 + 5 = _____

0 + 5 = _____   0 + 1 = _____

3 + 3 = _____   1 + 2 = _____

5 + 5 = _____   2 + 4 = _____

4 + 5 = _____   3 + 5 = _____

0 + 3 = _____   1 + 4 = _____

Addition Problems

5 + 5 = _____   4 + 0 = _____

4 + 3 = _____   5 + 1 = _____

4 + 4 = _____   0 + 0 = _____

1 + 1 = _____   3 + 2 = _____

5 + 2 = _____   5 + 5 = _____

2 + 0 = _____   3 + 1 = _____

Addition Problems

$5 + 3 =$ _____　　　$2 + 2 =$ _____

$2 + 1 =$ _____　　　$4 + 1 =$ _____

$4 + 2 =$ _____　　　$5 + 0 =$ _____

$5 + 5 =$ _____　　　$3 + 0 =$ _____

$3 + 3 =$ _____　　　$5 + 5 =$ _____

$1 + 0 =$ _____　　　$5 + 4 =$ _____

Ten Frame

Count the dots and write the quantity.

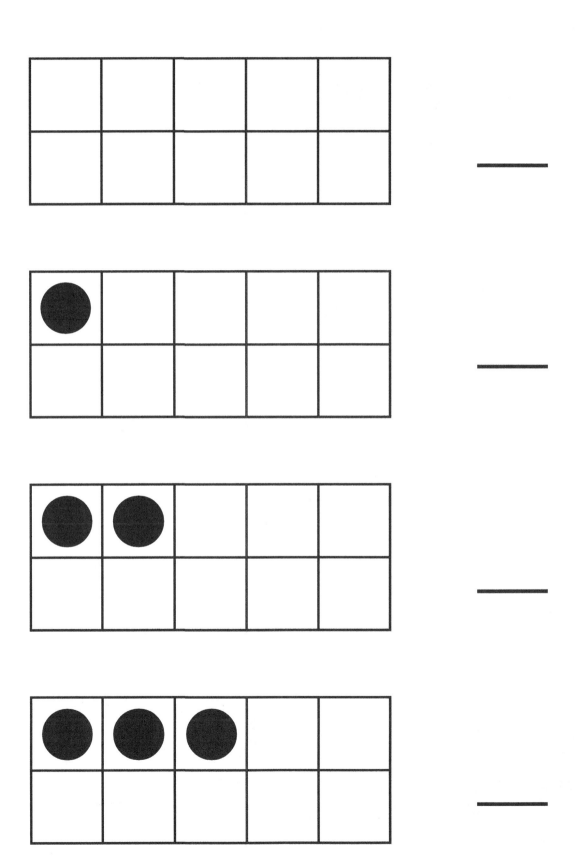

Ten Frame

Count and write the quantity.

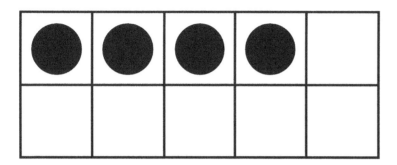

          ____

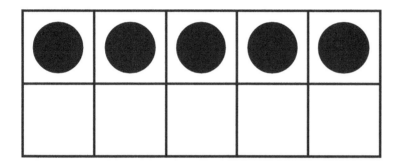

          ____

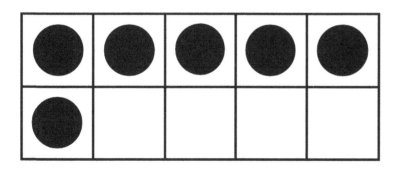

          ____

          ____

Ten Frame

Count and write the quantity.

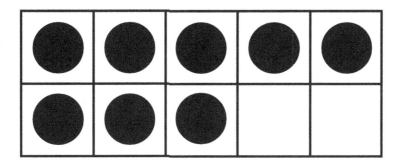

_____

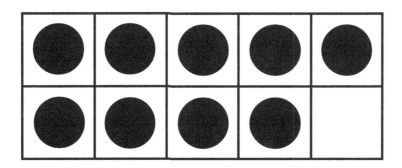

_____

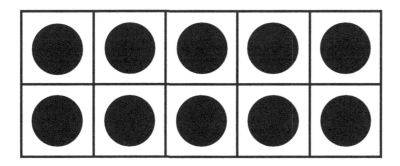

_____

Color the Ten Frame

Color the correct number of squares.

4

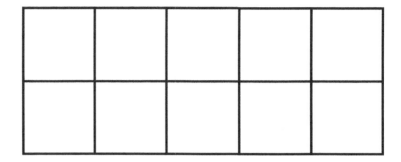

7

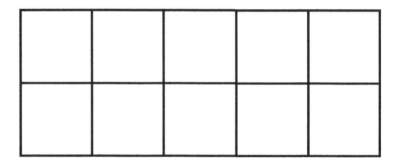

0

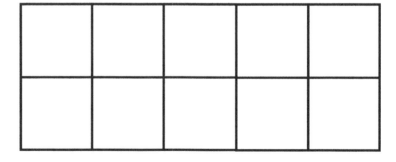

1

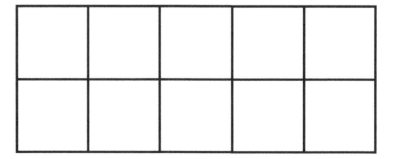

Color the Ten Frame

Color the correct number of squares.

3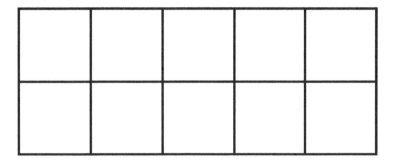

9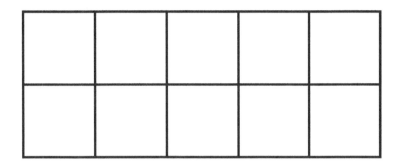

5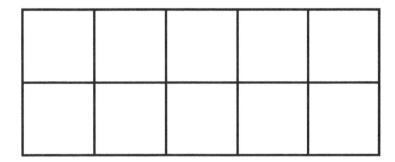

2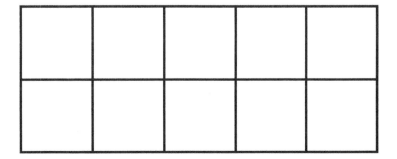

Color the Ten Frame

Color the correct number of squares.

10

8

3

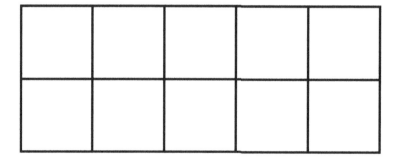

6

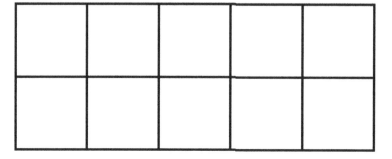

Eleven

Draw 11 dots in the ten frames. Make sure to completely fill the first ten frame before moving on to the second one.

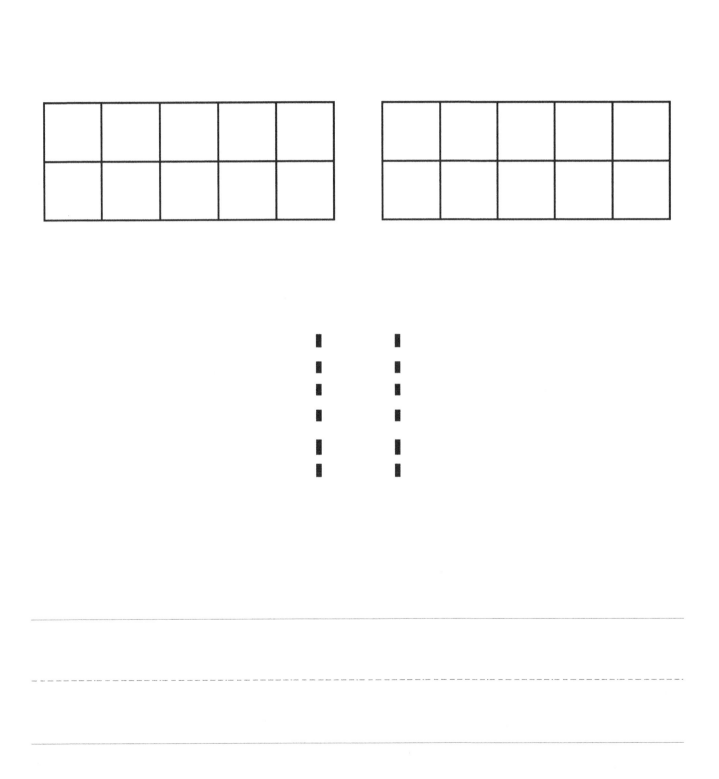

Number Quantity

Count and Color 11 frogs green.

11

Twelve

Draw 12 dots in the ten frames. Make sure to completely fill the first ten frame before moving on to the second one.

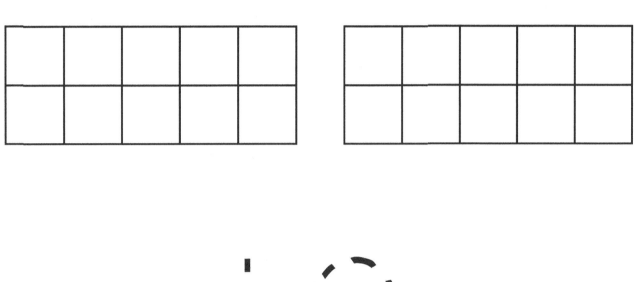

Number Quantity

Put an "x" on 12 apples.

12

Thirteen

Draw 13 dots in the ten frames. Make sure to completely fill the first ten frame before moving on to the second one.

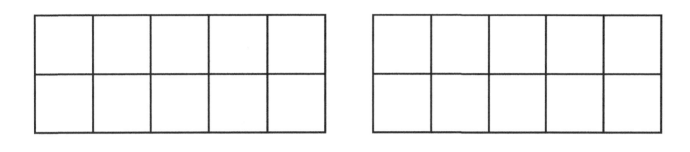

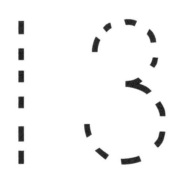

Number Quantity

Count and circle 13 fish.

13

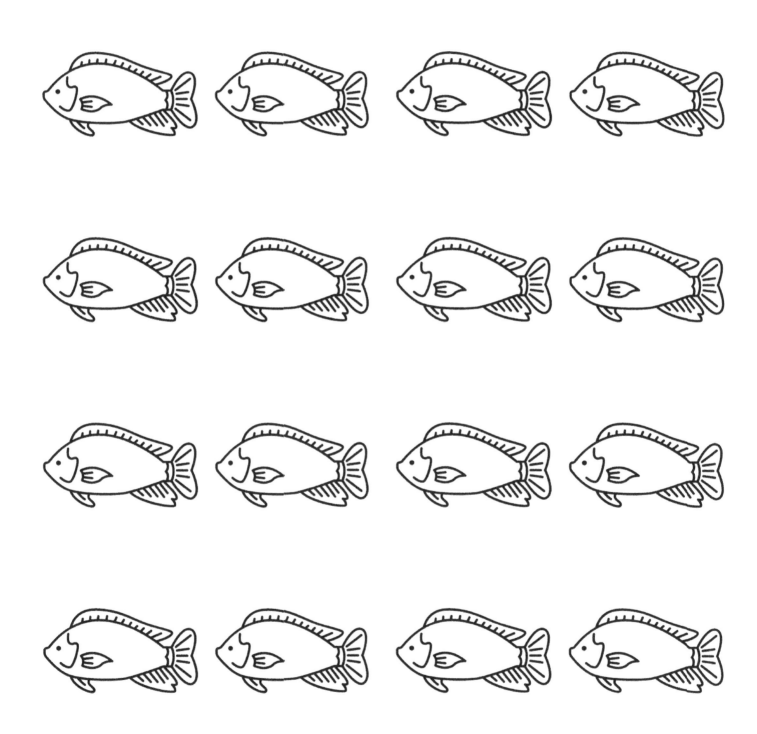

Fourteen

Draw 14 dots in the ten frames. Make sure to completely fill the first ten frame before moving on to the second one.

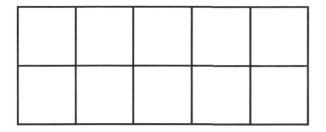 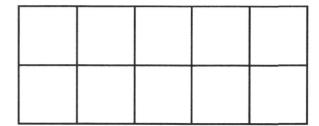

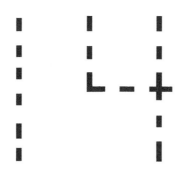

Number Quantity

Count and Color 14 crabs.

14

Fifteen

Draw 15 dots in the ten frames. Make sure to completely fill the first ten frame before moving on to the second one.

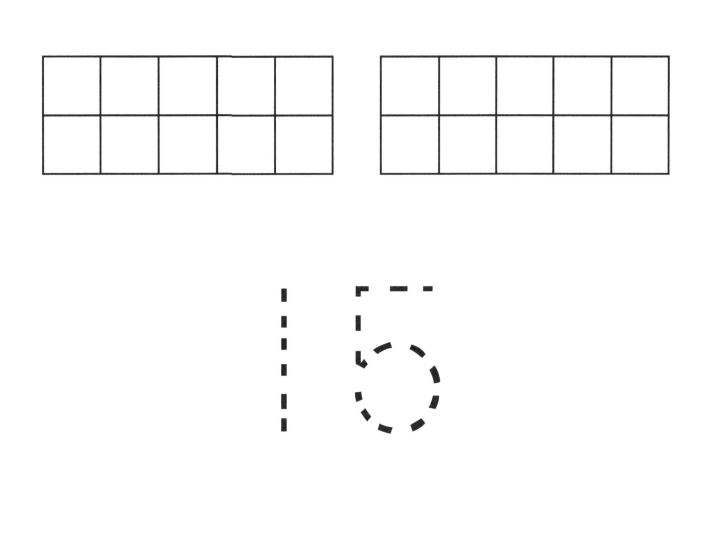

Number Quantity

Draw a square around 15 watermelon slices.

15

Sixteen

Draw 16 dots in the ten frames. Make sure to completely fill the first ten frame before moving on to the second one.

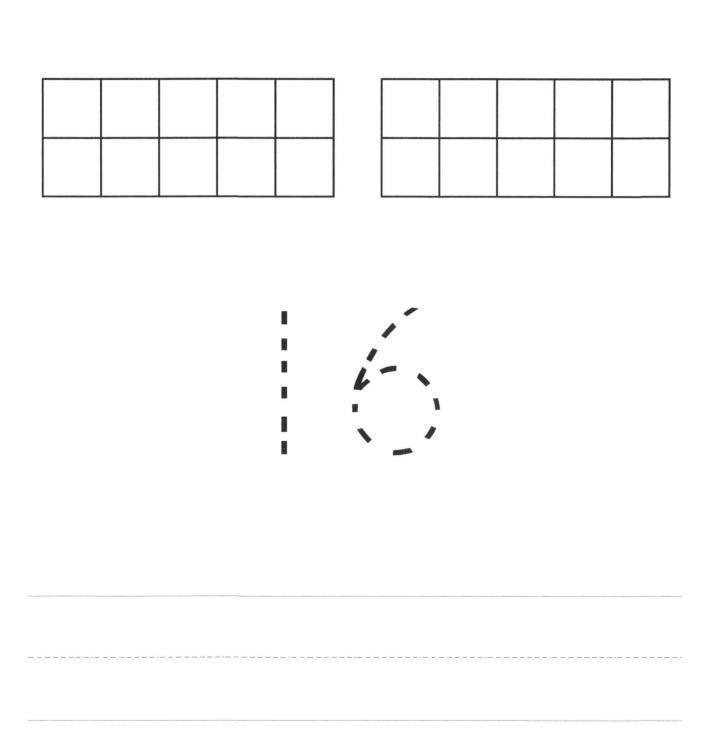

Number Quantity

Count and circle 16 feathers.

16

Seventeen

Draw 17 dots in the ten frames. Make sure to completely fill the first ten frame before moving on to the second one.

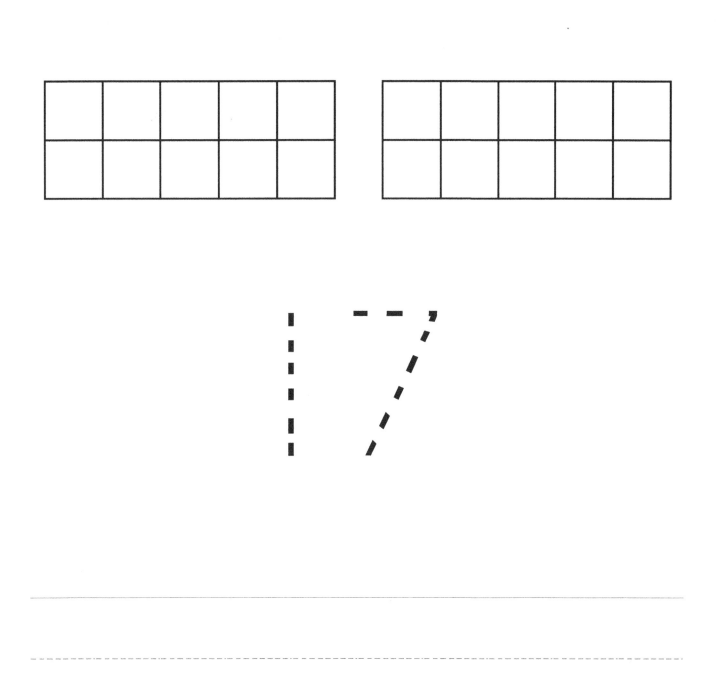

Number Quantity

Count and color 17 bees.

17

Eighteen

Draw 18 dots in the ten frames. Make sure to completely fill the first ten frame before moving on to the second one.

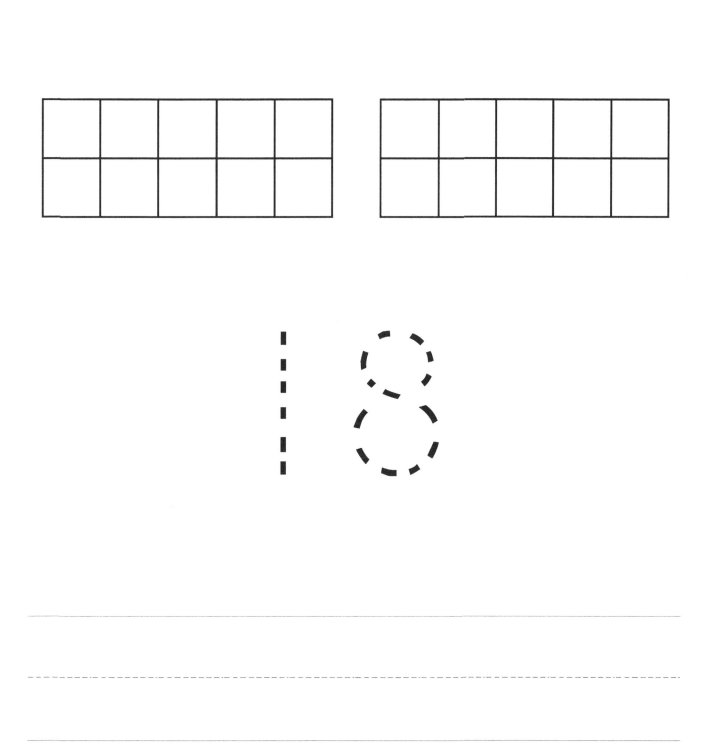

Number Quantity

Put an "x" on 18 lady bugs.

18

Nineteen

Draw 19 dots in the ten frames. Make sure to completely fill the first ten frame before moving on to the second one.

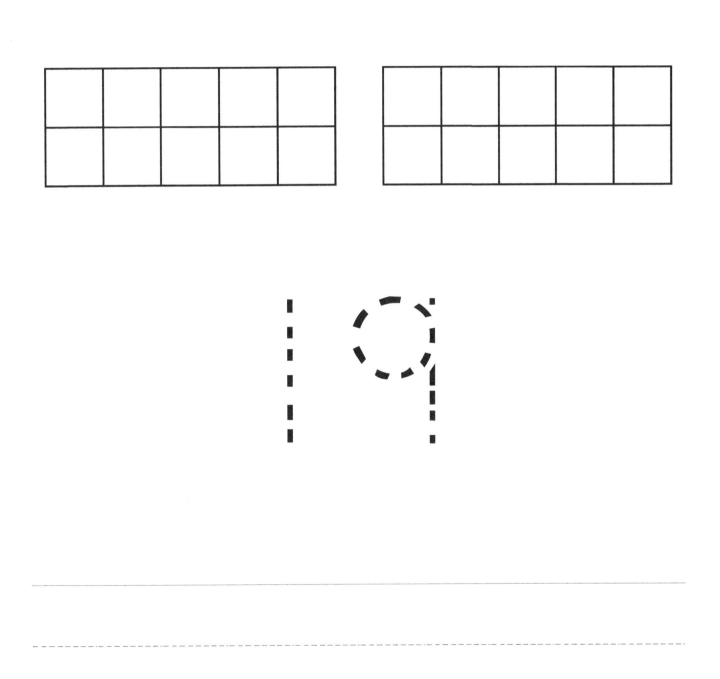

Number Quantity

Count and color 19 bananas.

19

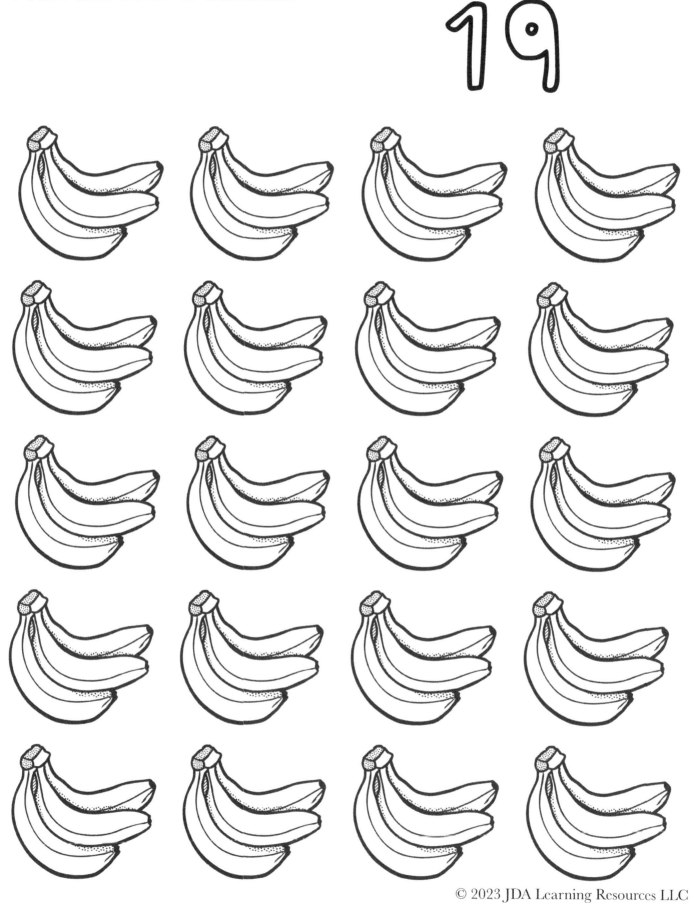

Twenty

Draw 20 dots in the ten frames. Make sure to completely fill the first ten frame before moving on to the second one.

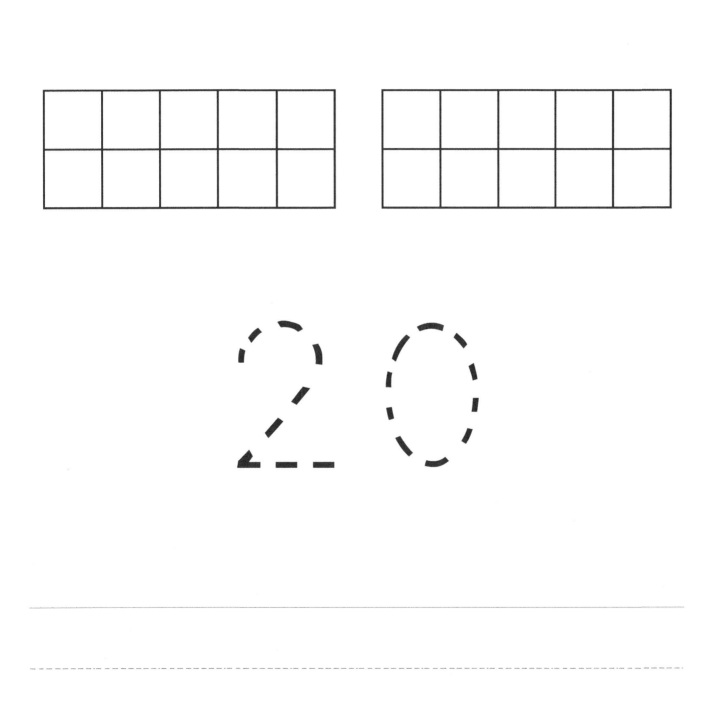

Number Quantity

Cross out 20 beach balls.

20

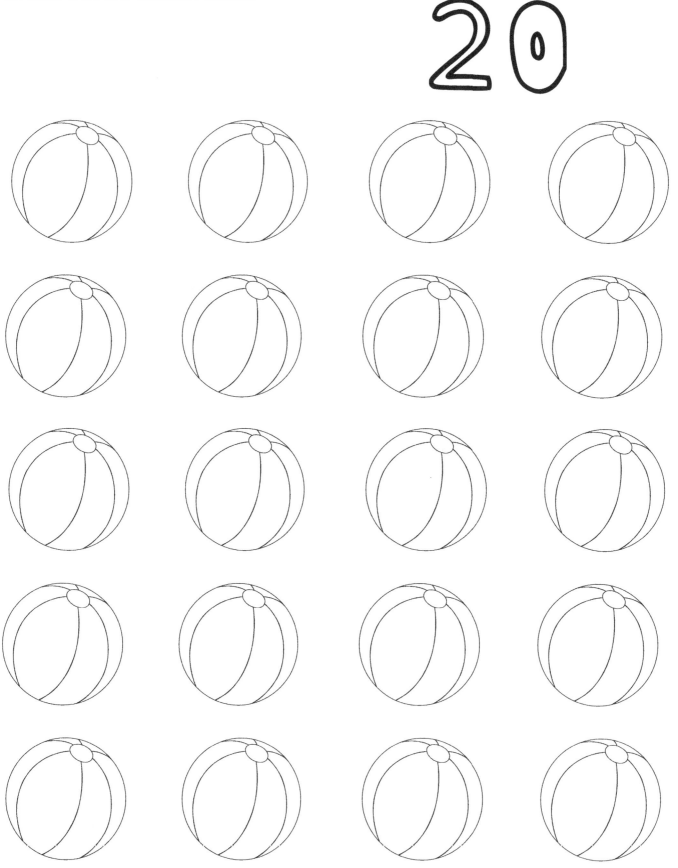

Matching

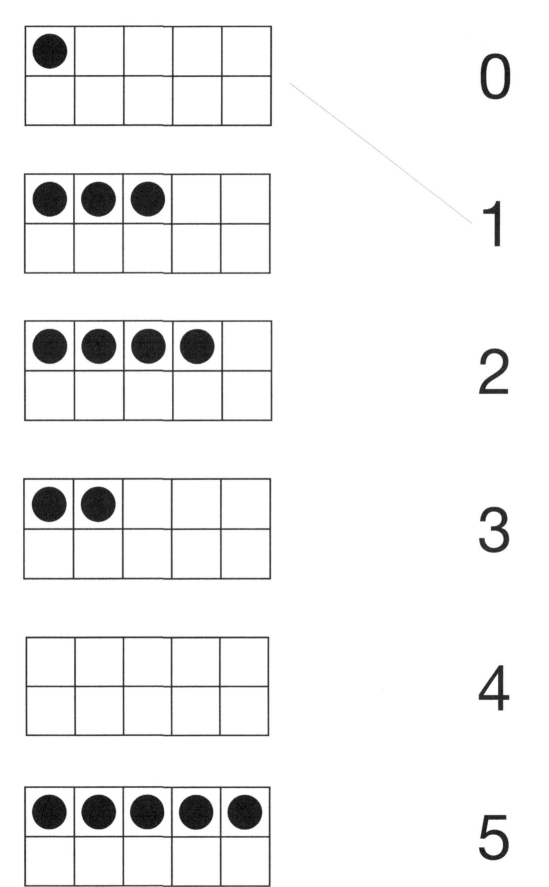

0

1

2

3

4

5

Matching

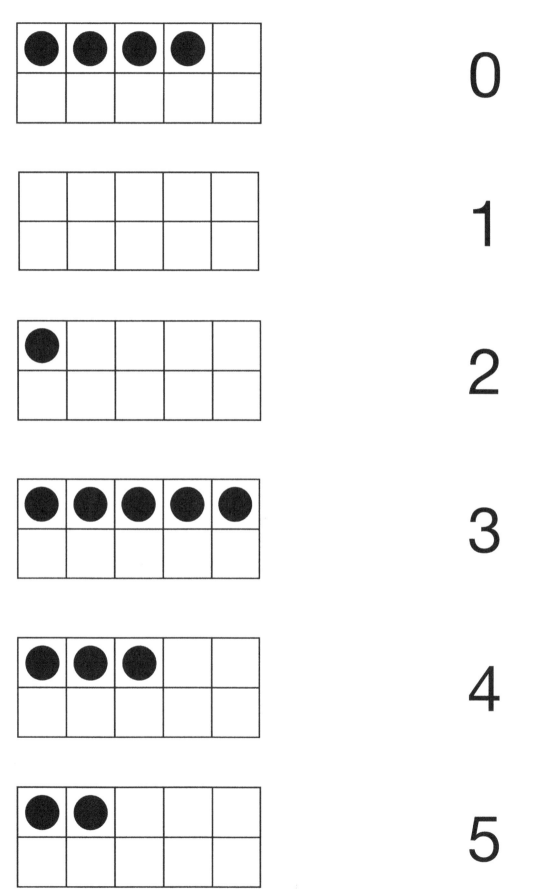

Matching

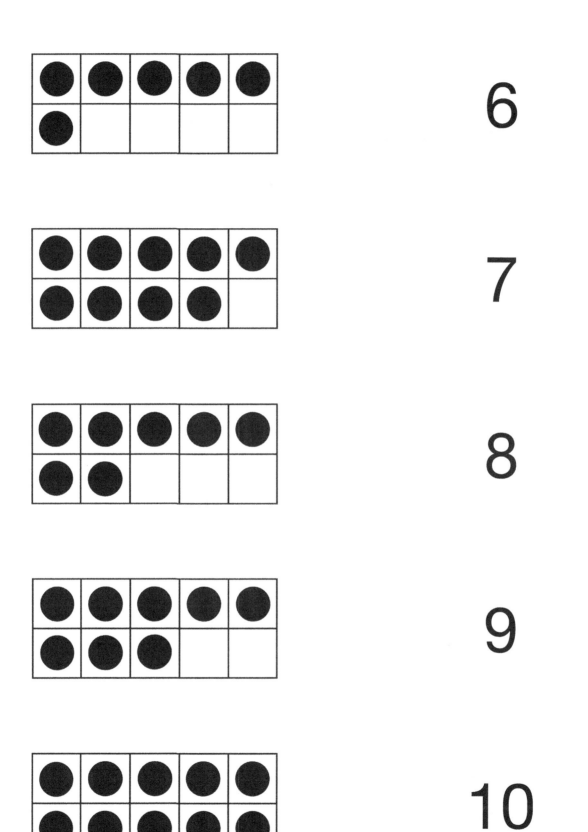

6

7

8

9

10

Matching

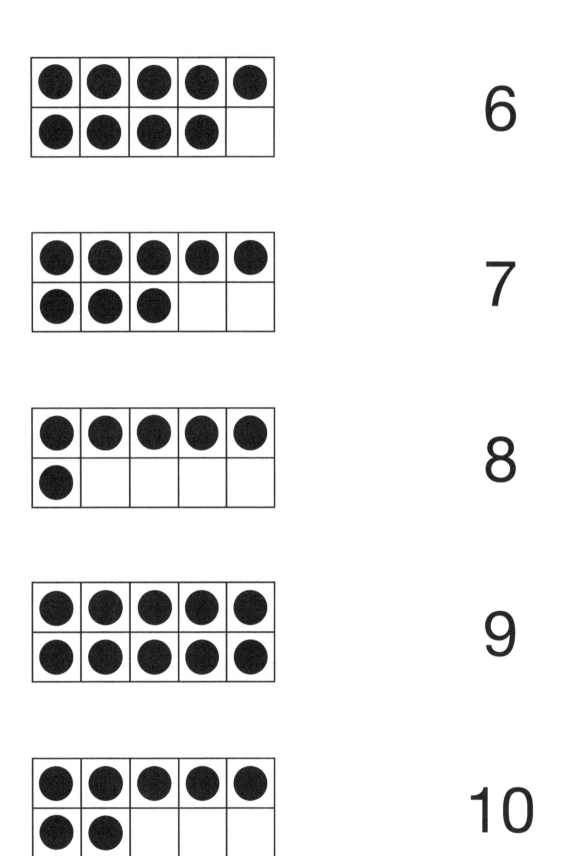

6

7

8

9

10

Matching

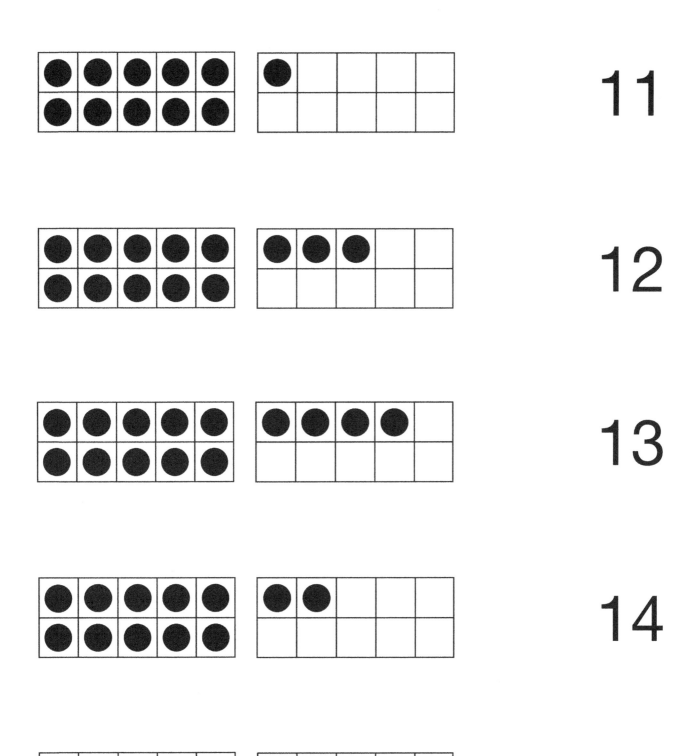

11

12

13

14

15

Matching

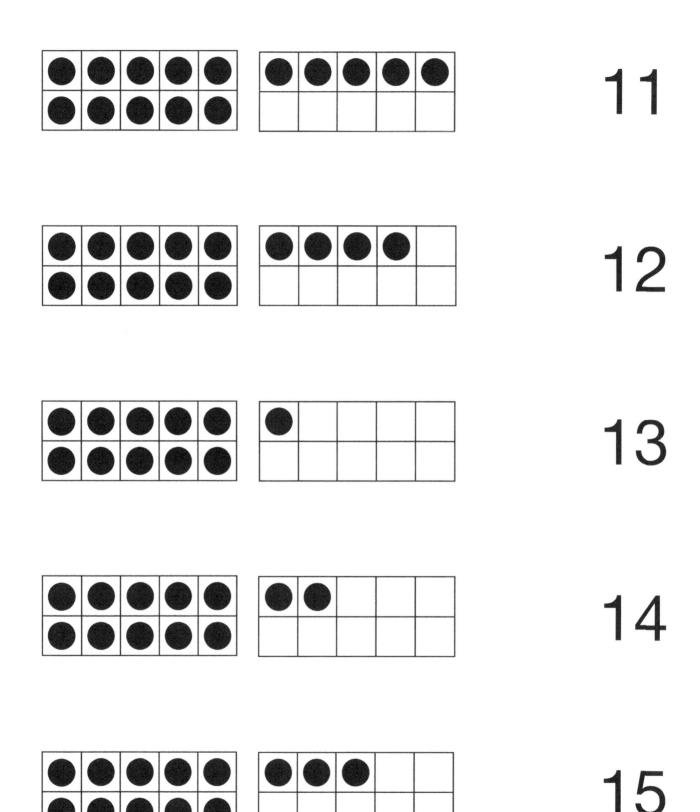

Matching

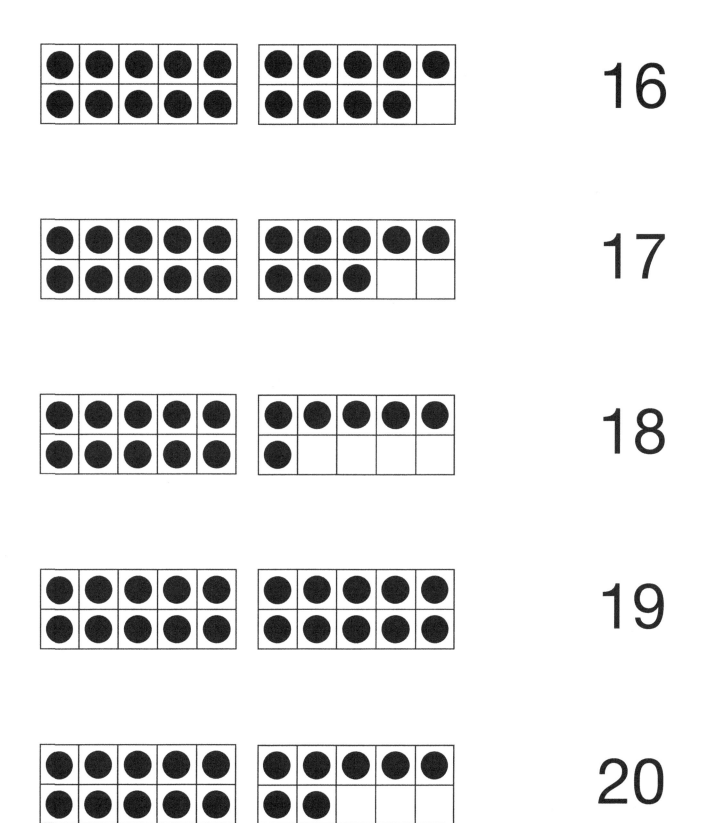

16

17

18

19

20

Matching

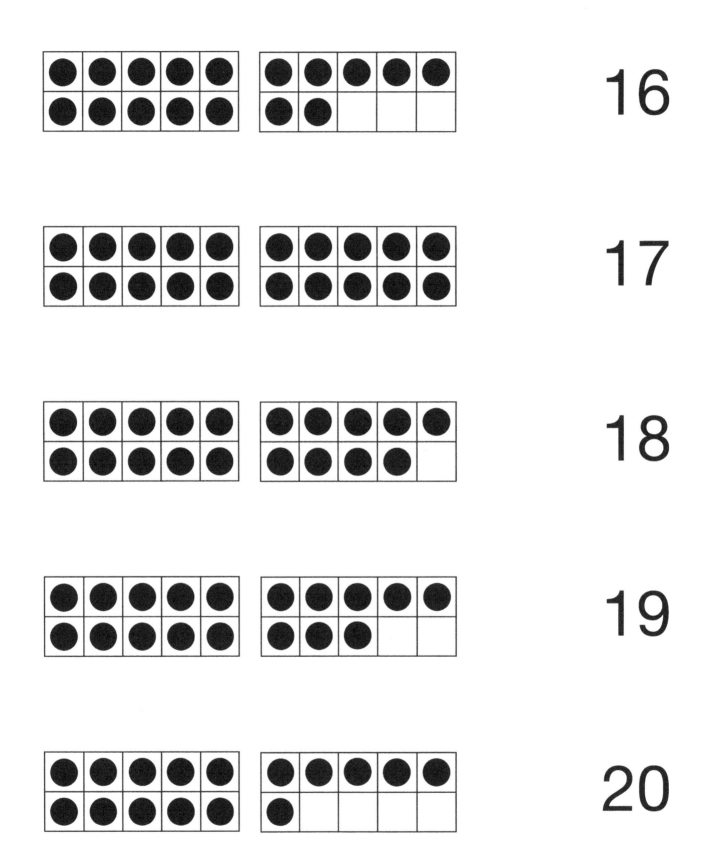

16

17

18

19

20

Color the Ten Frame

Color the correct number of squares.

14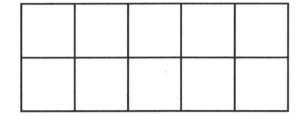

11

20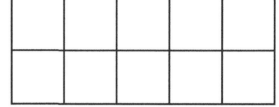

19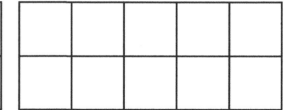

Color the Ten Frame

Color the correct number of squares.

12

17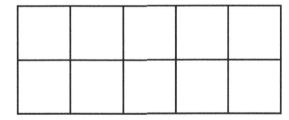

13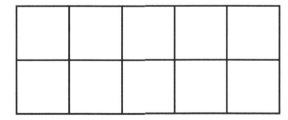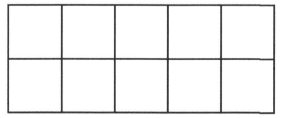

15

Color the Ten Frame

Color the correct number of squares.

20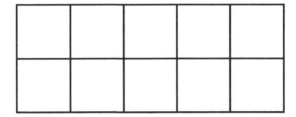

15

18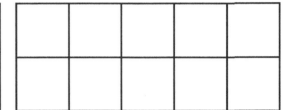

16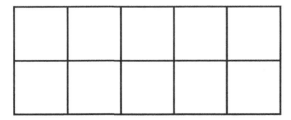

Adding with Ten Frames

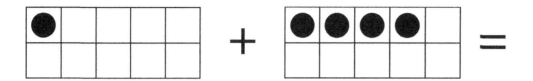

+ = ___

___ ___ ___

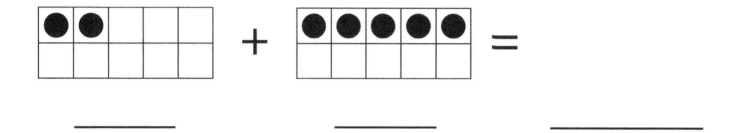

+ = ___

___ ___ ___

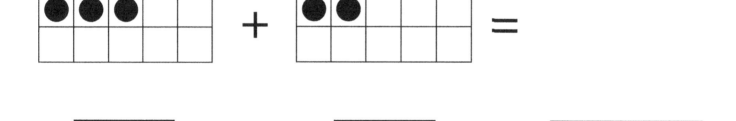

+ = ___

___ ___ ___

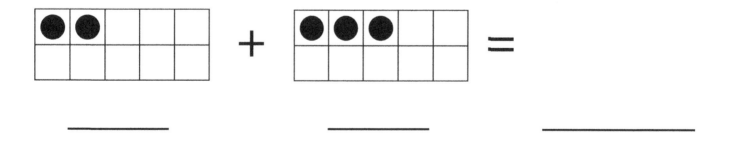

+ = ___

___ ___ ___

+ = ___

___ ___ ___

Adding with Ten Frames

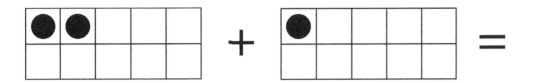

_____   _____   _____

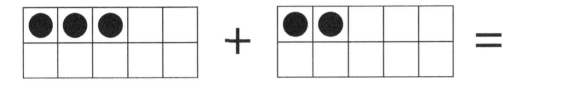

_____   _____   _____

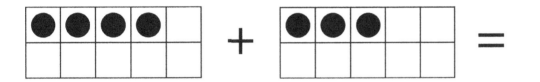

_____   _____   _____

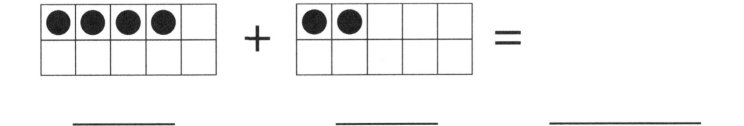

_____   _____   _____

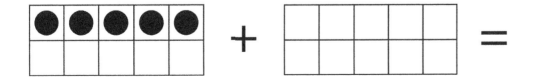

_____   _____   _____

Adding with Ten Frames

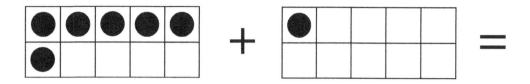

_____   _____   _____

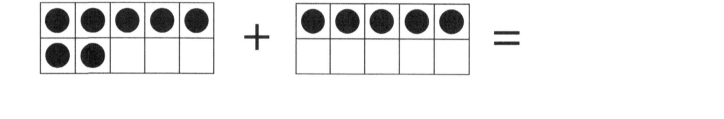

_____   _____   _____

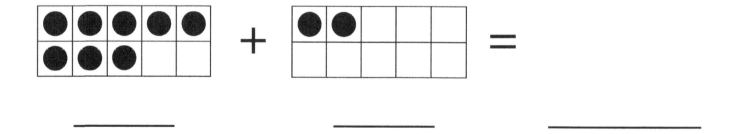

_____   _____   _____

_____   _____   _____

_____   _____

Adding with Ten Frames

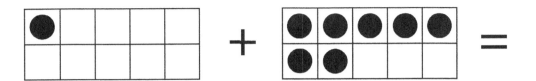 =

_____   _____   _____

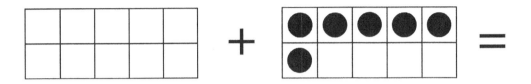 =

_____   _____   _____

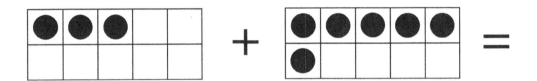 =

_____   _____   _____

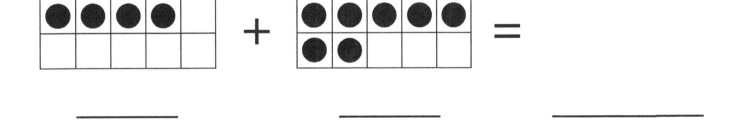 =

_____   _____   _____

_____   _____   _____

Adding with Ten Frames

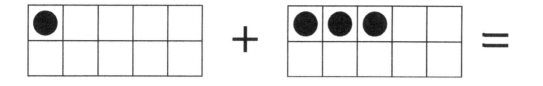

_____   +   _____   =   _____

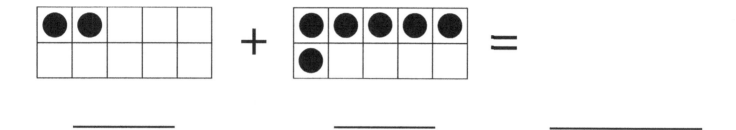

_____   +   _____   =   _____

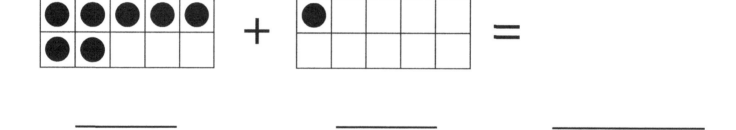

_____   +   _____   =   _____

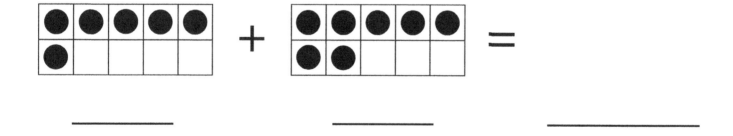

_____   +   _____   =   _____

_____   +   _____   =   _____

Adding with Ten Frames

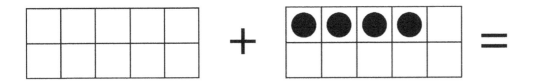

_____   _____   _____

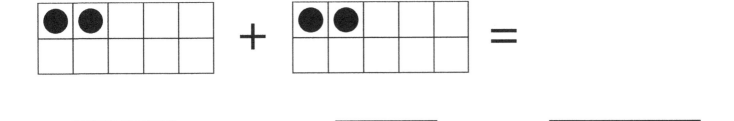

_____   _____   _____

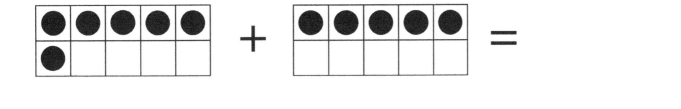

_____   _____   _____

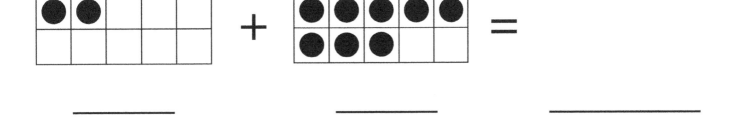

_____   _____   _____

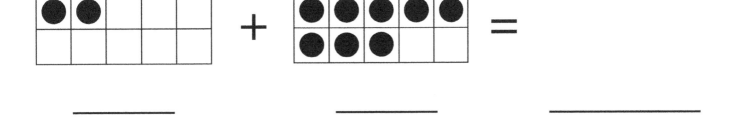

_____   _____   _____

Adding with Ten Frames

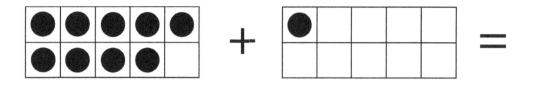

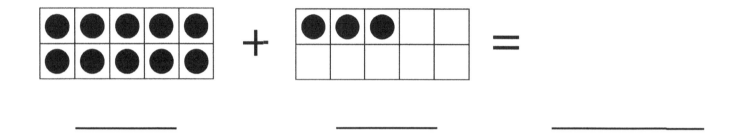

_____   _____   _____

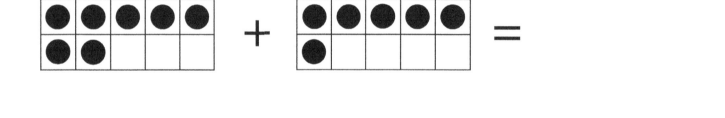

_____   _____   _____

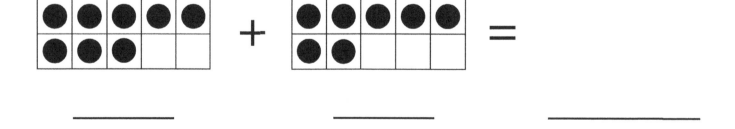

_____   _____   _____

_____   _____   _____

Adding with Ten Frames

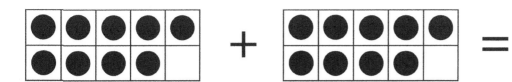

_____   _____   _____

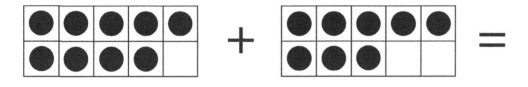

_____   _____   _____

_____   _____   _____

_____   _____   _____

_____   _____   _____

Adding with Ten Frames

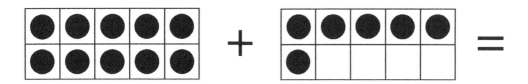 =

_____   _____   _____

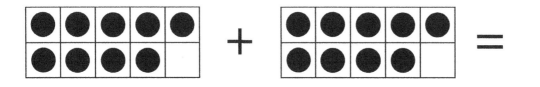 =

_____   _____   _____

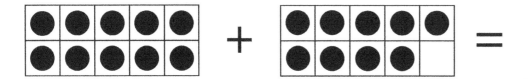 =

_____   _____   _____

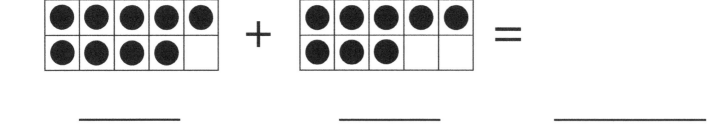 =

_____   _____   _____

  =

_____   _____   _____

Adding with Ten Frames

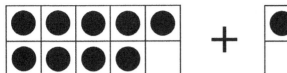   +   =

_____   _____   _____

   +   =

_____   _____   _____

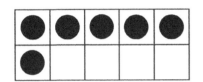   +   =

_____   _____   _____

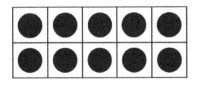   +   =

_____   _____   _____

 +   =

_____   _____   _____

Counting On Example

Take the larger number and place it in your mind.
Make the smaller number using your fingers.
Count your fingers starting from the next number.

6 + 2 = ?

Counting On Example

7 + 3 = ?

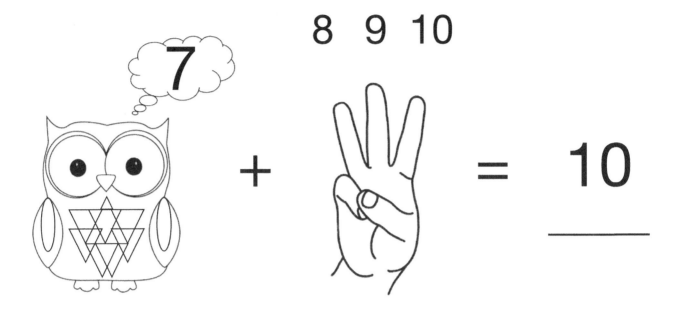

8 + 4 = ?

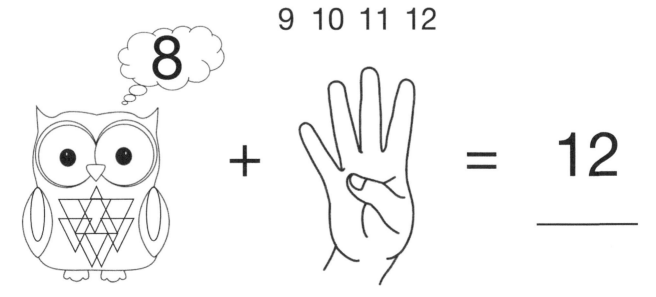

Counting On Problems

Place the larger number in the owl's bubble and count on.

7 + 2 = ?   +   = _____

5 + 4 = ?   +   = _____

6 + 1 = ?   +   = _____

8 + 5 = ?   +   = _____

Counting On Problems

Place the larger number in the owl's bubble and count on.

1 + 9 = ?

3 + 7 = ?

4 + 6 = ?

2 + 8 = ?

Counting On Problems

Place the larger number in the owl's bubble and count on.

12 + 3 = ?   +   = _____

8 + 6 = ?   +   = _____

11 + 5 = ?   +   = _____

10 + 7 = ?   +   = _____

Counting On Problems

Place the larger number in the owl's bubble and count on.

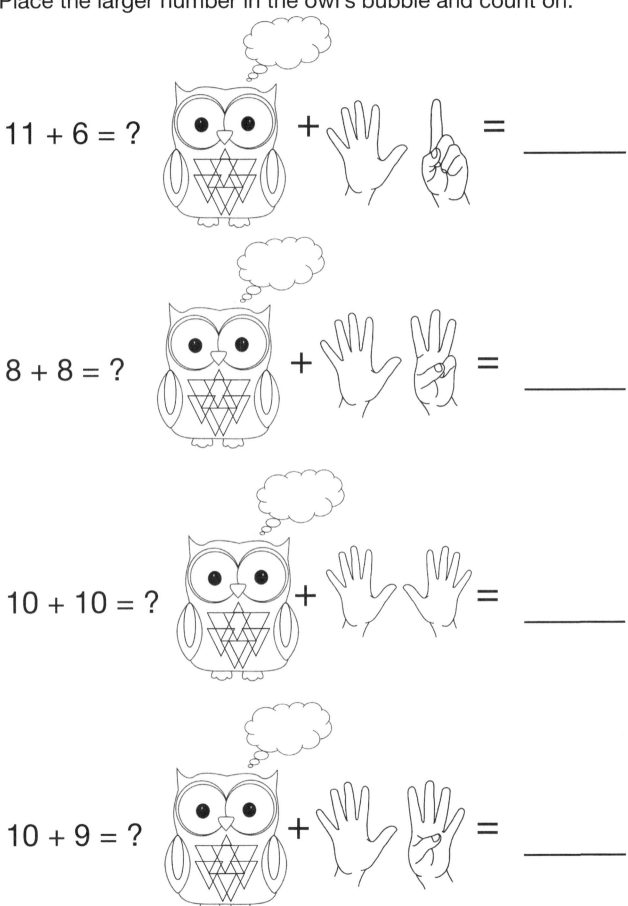

$11 + 6 = ?$ +   = _____

$8 + 8 = ?$ +   = _____

$10 + 10 = ?$ +   = _____

$10 + 9 = ?$ +   = _____

Counting On Problems

Place the larger number in the owl's bubble and count on.

3 + 3 = ?   + = _____

11 + 3 = ?   + = _____

4 + 4 = ?   + = _____

6 + 8 = ?   + = _____

Counting On Problems
Place the larger number in the owl's bubble and count on.

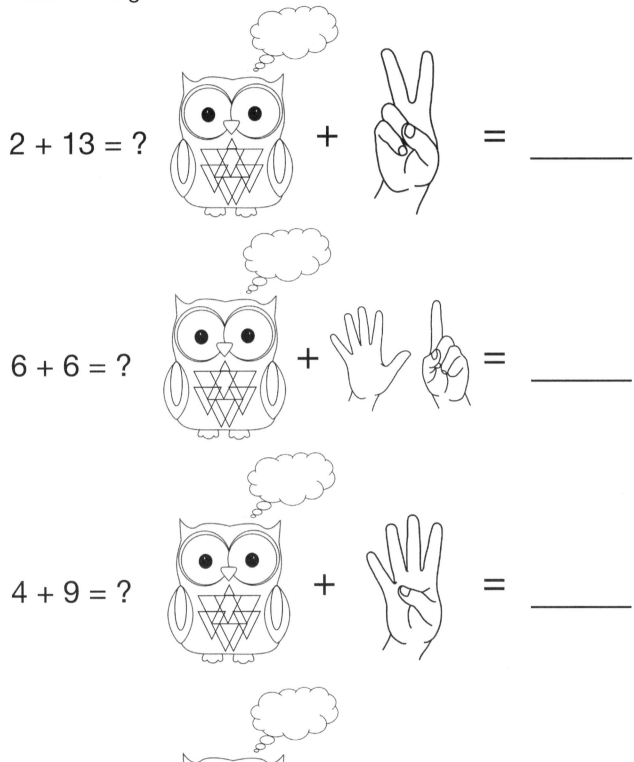

2 + 13 = ?   +   = _____

6 + 6 = ?   +   = _____

4 + 9 = ?   +   = _____

7 + 7 = ?

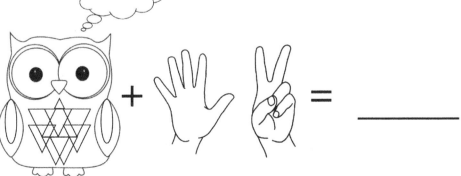

+   = _____

Counting On Problems

Place the larger number in the owl's bubble and count on.

$14 + 6 = ?$ +   = _____

$7 + 3 = ?$ +   = _____

$15 + 5 = ?$ +   = _____

$9 + 8 = ?$ +   = _____

Counting On Problems

Place the larger number in the owl's bubble and count on.

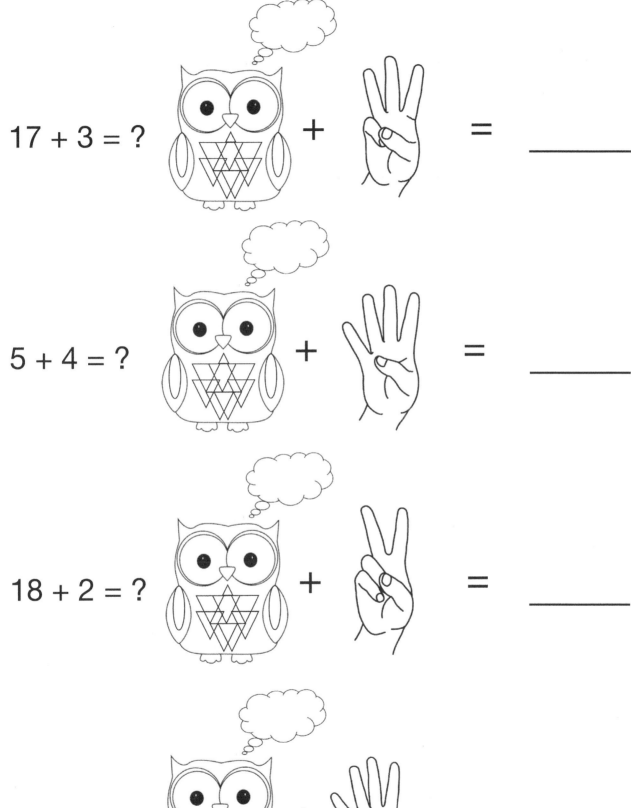

17 + 3 = ?   +   = ____

5 + 4 = ?   +   = ____

18 + 2 = ?   +   = ____

6 + 5 = ?   +   = ____

Counting On Problems

Place the larger number in the owl's bubble and count on.

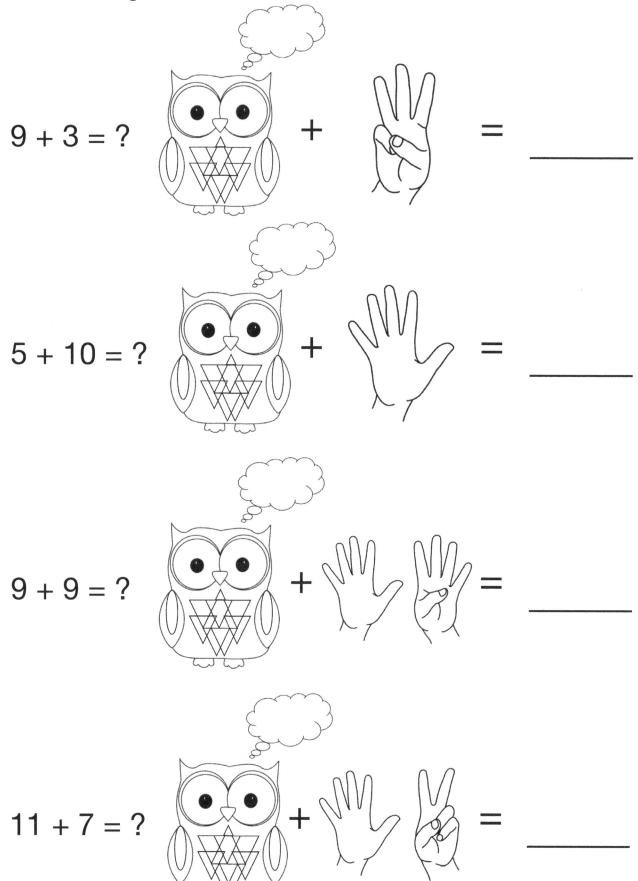

9 + 3 = ?   +   = _____

5 + 10 = ?   +   = _____

9 + 9 = ?   +   = _____

11 + 7 = ?   +   = _____

Counting On Problems

Place the larger number in the owl's bubble and count on.

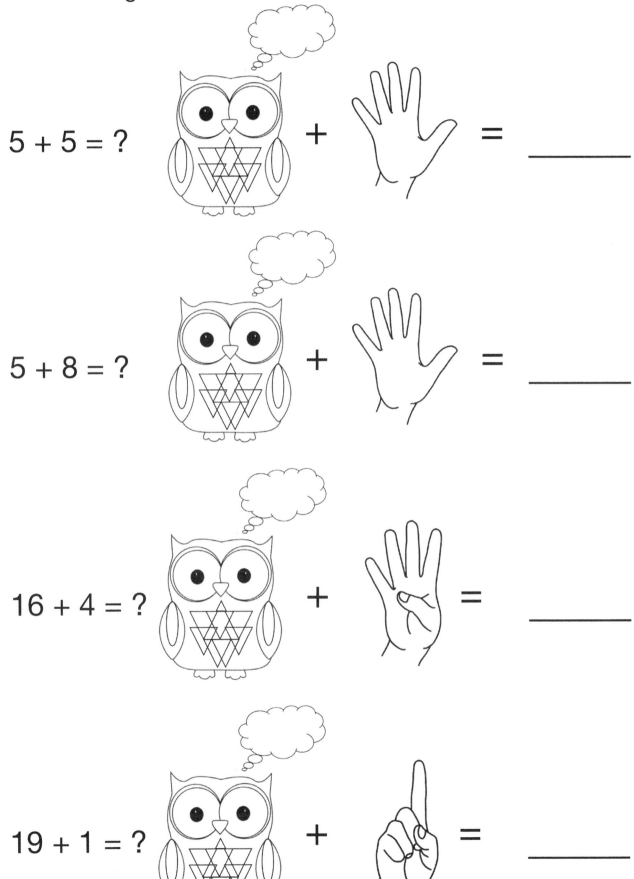

$5 + 5 = ?$ + = _____

$5 + 8 = ?$ + = _____

$16 + 4 = ?$ + = _____

$19 + 1 = ?$ + = _____

What Makes 10

Count on to find the missing number.

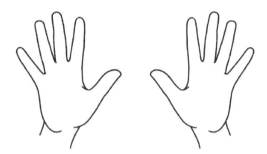 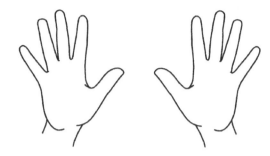

3 + _____ = 10   2 + _____ = 10

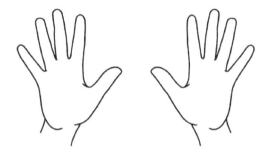

4 + _____ = 10   1 + _____ = 10

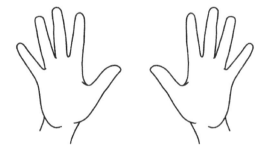

5 + _____ = 10   6 + _____ = 10

What Makes 10

Count on to find the missing number.

7 + _____ = 10

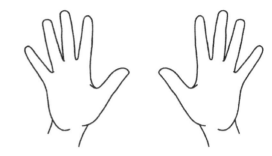

1 + _____ = 10

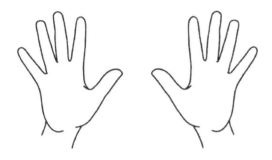

9 + _____ = 10

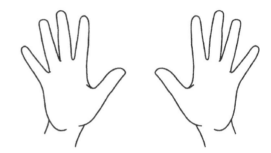

8 + _____ = 10

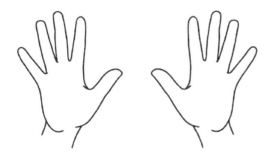

4 + _____ = 10

2 + _____ = 10

What Makes 10

Count on to find the missing number.

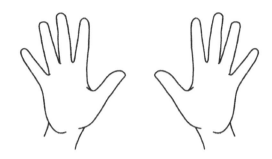 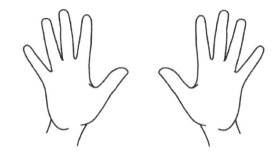

_____ + 3 = 10   _____ + 2 = 10

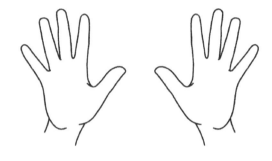

_____ + 4 = 10   _____ + 1 = 10

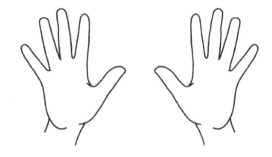

_____ + 5 = 10   _____ + 6 = 10

What Makes 10

Count on to find the missing number.

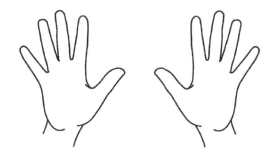

_____ + 7 = 10

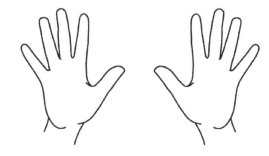

_____ + 1 = 10

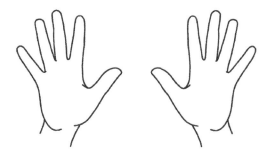

9 + _____ = 10

8 + _____ = 10

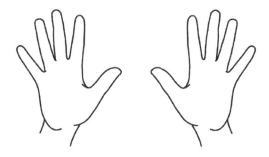

4 + _____ = 10

2 + _____ = 10

What Makes 10

How many are missing?
Count on to find the missing number

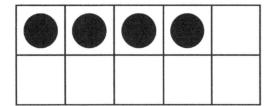

4 + _____ = 10

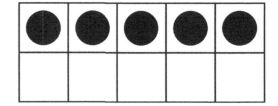

5 + _____ = 10

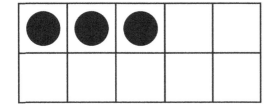

3 + _____ = 10

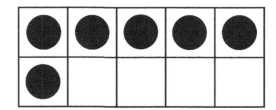

6 + _____ = 10

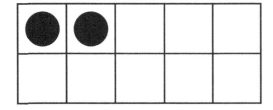

2 + _____ = 10

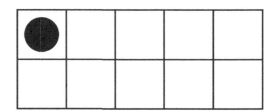

1 + _____ = 10

What Makes 10

How many are missing?
Count on to find the missing number

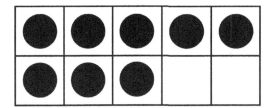

$8 +$ _____ $= 10$

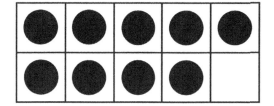

$9 +$ _____ $= 10$

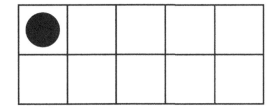

$1 +$ _____ $= 10$

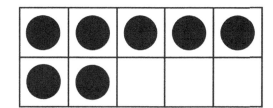

$7 +$ _____ $= 10$

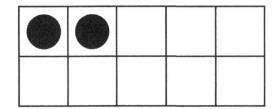

$2 +$ _____ $= 10$

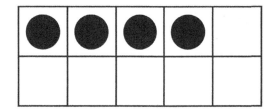

$4 +$ _____ $= 10$

What Makes 10

How many are missing?
Count on to find the missing number

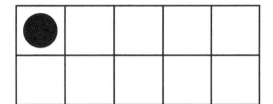

_____ + 1 = 10

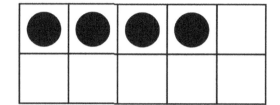

_____ + 4 = 10

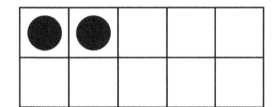

_____ + 2 = 10

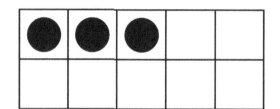

_____ + 3 = 10

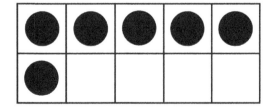

_____ + 6 = 10

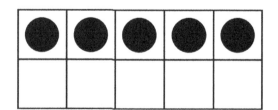

_____ + 5 = 10

What Makes 10

How many are missing?
Count on to find the missing number

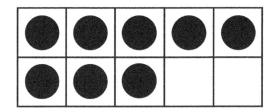

_____ + 8 = 10

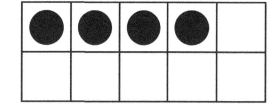

_____ + 4 = 10

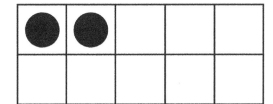

_____ + 2 = 10

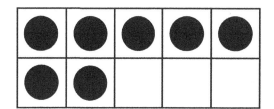

_____ + 7 = 10

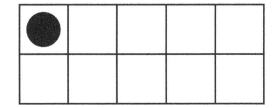

_____ + 1 = 10

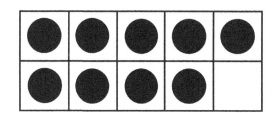

_____ + 9 = 10

Doubles Matching

1 + 1 = 2   2 + 2 = 4

2 + 2 = 4   5 + 5 = 10

3 + 3 = 6   1 + 1 = 2

4 + 4 = 8   3 + 3 = 6

5 + 5 = 10   4 + 4 = 8

Doubles Matching

6 + 6 = 12

8 + 8 = 16

7 + 7 = 14

10 + 10 = 20

8 + 8 = 16

6 + 6 = 12

9 + 9 = 18

9 + 9 = 18

10 + 10 = 20

7 + 7 = 14

Doubles Fill in the Blank

Use the box to find the answer.

$2 + 2 = 4$	$4 + 4 = 8$	$6 + 6 = 12$
$5 + 5 = 10$	$1 + 1 = 2$	$3 + 3 = 6$

$1 + 1 =$ _____ $\qquad\qquad$ $2 + 2 =$ _____

$3 + 3 =$ _____ $\qquad\qquad$ $4 + 4 =$ _____

$5 + 5 =$ _____ $\qquad\qquad$ $6 + 6 =$ _____

Doubles Fill in the Blank

Use the box to find the answer.

$$7 + 7 = 14 \qquad 9 + 9 = 18 \qquad 6 + 6 = 12$$

$$5 + 5 = 10 \qquad 10 + 10 = 20 \qquad 8 + 8 = 16$$

$10 + 10 = $ _____ $\qquad\qquad$ $8 + 8 = $ _____

$7 + 7 = $ _____ $\qquad\qquad$ $9 + 9 = $ _____

$5 + 5 = $ _____ $\qquad\qquad$ $6 + 6 = $ _____

Double Memory Work

The double of 4 is _____

The double of 7 is _____

The double of 1 is _____

The double of 8 is _____

The double of 5 is _____

Double Memory Work

The double of 6 is _____

The double of 2 is _____

The double of 10 is _____

The double of 3 is _____

The double of 9 is _____

Doubles Problems

6 + 6 = _____

1 + 1 = _____

4 + 4 = _____

8 + 8 = _____

5 + 5 = _____

2 + 2 = _____

9 + 9 = _____

10 + 10 = _____

3 + 3 = _____

9 + 9 = _____

7 + 7 = _____

0 + 0 = _____

Doubles Problems

9 + 9 = _____

3 + 3 = _____

10 + 10 = _____

7 + 7 = _____

0 + 0 = _____

2 + 2 = _____

9 + 9 = _____

8 + 8 = _____

6 + 6 = _____

4 + 4 = _____

1 + 1 = _____

5 + 5 = _____

Doubles + 1 Addition Technique

Take the smaller number.
Double it.
Then Add 1.

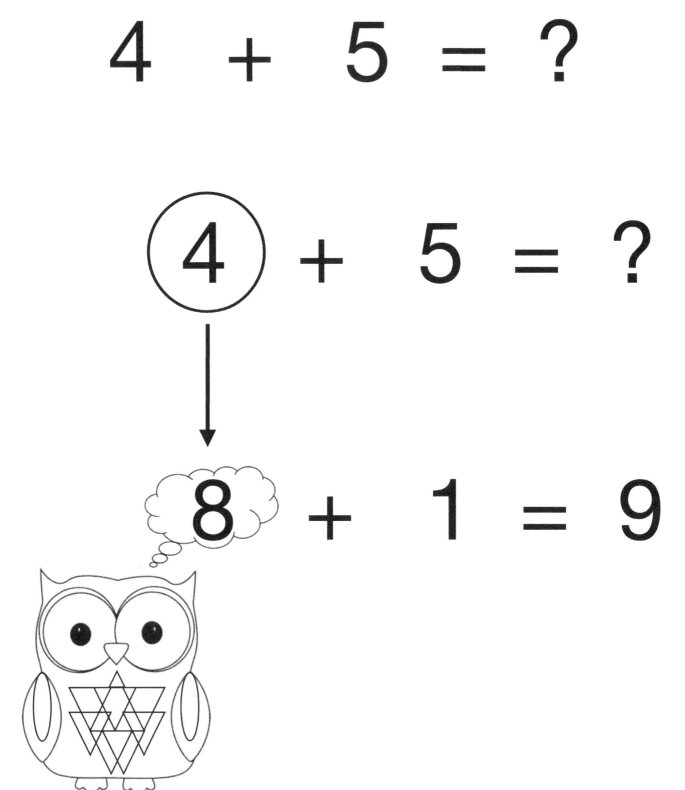

Doubles + 1

Double the smaller number, write it in the the owl's bubble, and add 1.

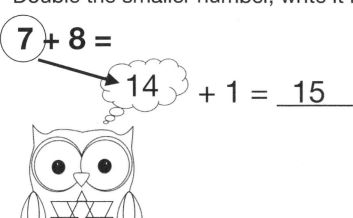

⑦ + 8 =

14 + 1 = __15__

8 + 7 =

___ + 1 = _____

5 + 4 =

___ + 1 = _____

10 + 9 =

___ + 1 = _____

5 + 6 =

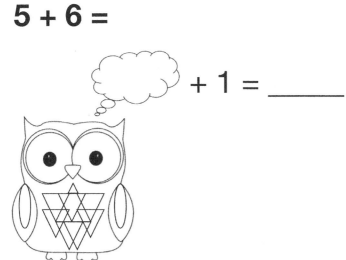

___ + 1 = _____

6 + 5 =

___ + 1 = _____

Doubles + 1

Double the smaller number, write it in the the owl's bubble, and add 1.

1 + 2 =

 + 1 = _____

8 + 9 =

 + 1 = _____

3 + 2 =

 + 1 = _____

2 + 1 =

 + 1 = _____

3 + 4 =

 + 1 = _____

9 + 8 =

 + 1 = _____

Doubles + 1

Double the smaller number, write it in the the owl's bubble, and add 1.

6 + 7 =

 + 1 = _____

4 + 3 =

 + 1 = _____

3 + 2 =

 + 1 = _____

9 + 10 =

 + 1 = _____

4 + 5 =

 + 1 = _____

7 + 6 =

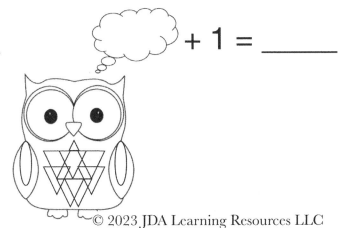 + 1 = _____

Doubles + 1 Matching

Double each number and add 1.

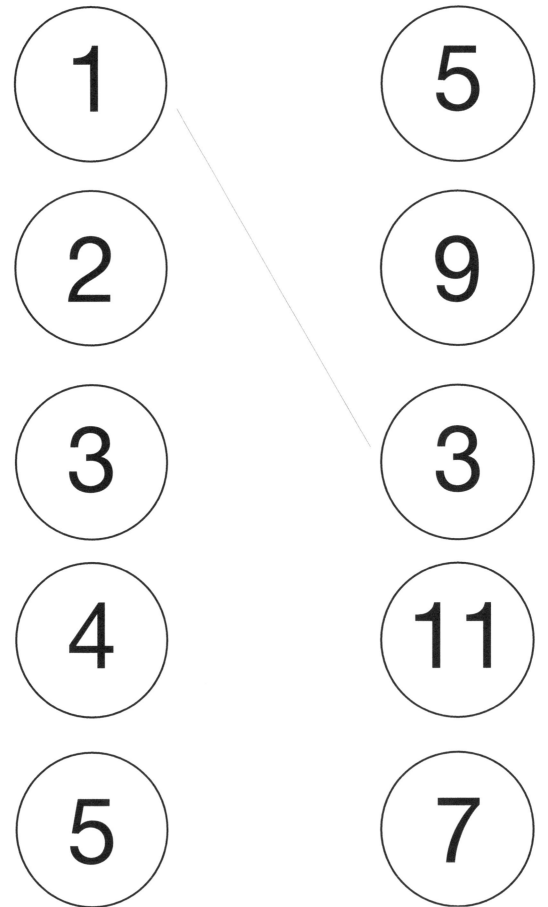

Doubles + 1 Matching
Double each number and add 1.

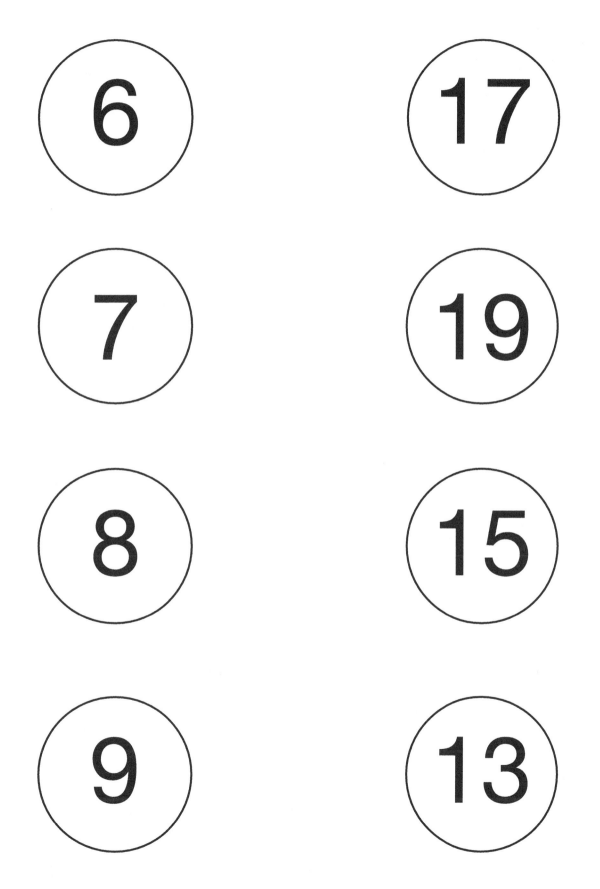

6   17

7   19

8   15

9   13

Doubles +1 Problems

6 + 7 = _____

1 + 2 = _____

4 + 5 = _____

8 + 9 = _____

5 + 6 = _____

2 + 3 = _____

9 + 10 = _____

6 + 5 = _____

3 + 4 = _____

10 + 9 = _____

8 + 7 = _____

1 + 0 = _____

Doubles +1 Problems

9 + 8 = _____

4 + 3 = _____

2 + 1 = _____

8 + 7 = _____

0 + 1 = _____

3 + 2 = _____

9 + 10 = _____

8 + 9 = _____

7 + 6 = _____

3 + 4 = _____

1 + 2 = _____

6 + 5 = _____

Mixed Problems

Solve the problems using all of the previously learned methods.

$20 + 0 =$ _____

$6 + 0 =$ _____

$3 + 16 =$ _____

$3 + 17 =$ _____

$6 + 14 =$ _____

$17 + 1 =$ _____

$12 + 1 =$ _____

$5 + 9 =$ _____

$8 + 2 =$ _____

$6 + 13 =$ _____

Mixed Problems

Solve the problems using all of the previously learned methods.

$7 + 13 =$ _____ $4 + 2 =$ _____

$6 + 6 =$ _____ $5 + 5 =$ _____

$10 + 4 =$ _____ $1 + 11 =$ _____

$11 + 6 =$ _____ $10 + 0 =$ _____

$9 + 11 =$ _____ $7 + 3 =$ _____

Mixed Problems

Solve the problems using all of the previously learned methods.

$2 + 2 =$ _____

$5 + 2 =$ _____

$0 + 10 =$ _____

$8 + 12 =$ _____

$14 + 4 =$ _____

$7 + 1 =$ _____

$3 + 9 =$ _____

$3 + 3 =$ _____

$7 + 13 =$ _____

$3 + 16 =$ _____

Mixed Problems

Solve the problems using all of the previously learned methods.

$6 + 10 =$ _____ $8 + 6 =$ _____

$5 + 8 =$ _____ $5 + 7 =$ _____

$14 + 3 =$ _____ $3 + 0 =$ _____

$0 + 5 =$ _____ $6 + 0 =$ _____

$3 + 11 =$ _____ $18 + 1 =$ _____

Mixed Problems

Solve the problems using all of the previously learned methods.

$2 + 2 =$ _____

$12 + 0 =$ _____

$6 + 14 =$ _____

$6 + 12 =$ _____

$8 + 10 =$ _____

$11 + 2 =$ _____

$3 + 12 =$ _____

$1 + 17 =$ _____

$6 + 1 =$ _____

$0 + 20 =$ _____

Mixed Problems

Solve the problems using all of the previously learned methods.

$10 + 10 = \underline{\hspace{2cm}}$ $8 + 12 = \underline{\hspace{2cm}}$

$6 + 8 = \underline{\hspace{2cm}}$ $0 + 6 = \underline{\hspace{2cm}}$

$15 + 0 = \underline{\hspace{2cm}}$ $0 + 10 = \underline{\hspace{2cm}}$

$10 + 5 = \underline{\hspace{2cm}}$ $8 + 8 = \underline{\hspace{2cm}}$

$2 + 13 = \underline{\hspace{2cm}}$ $15 + 1 = \underline{\hspace{2cm}}$

Mixed Problems

Solve the problems using all of the previously learned methods.

$10 + 10 =$ _____

$4 + 0 =$ _____

$6 + 1 =$ _____

$1 + 19 =$ _____

$7 + 13 =$ _____

$16 + 4 =$ _____

$3 + 9 =$ _____

$2 + 4 =$ _____

$3 + 14 =$ _____

$16 + 3 =$ _____

Mixed Problems

Solve the problems using all of the previously learned methods.

$0 + 15 = $ _____

$17 + 0 = $ _____

$9 + 11 = $ _____

$9 + 4 = $ _____

$4 + 15 = $ _____

$7 + 5 = $ _____

$7 + 7 = $ _____

$5 + 5 = $ _____

$5 + 15 = $ _____

$3 + 12 = $ _____

Mixed Problems

Solve the problems using all of the previously learned methods.

$4 + 3 =$ _____

$7 + 8 =$ _____

$8 + 8 =$ _____

$13 + 1 =$ _____

$15 + 0 =$ _____

$0 + 4 =$ _____

$2 + 3 =$ _____

$4 + 9 =$ _____

$16 + 2 =$ _____

$9 + 5 =$ _____

Mixed Problems

Solve the problems using all of the previously learned methods.

$1 + 18 =$ _____

$4 + 4 =$ _____

$1 + 0 =$ _____

$3 + 10 =$ _____

$4 + 16 =$ _____

$5 + 8 =$ _____

$5 + 13 =$ _____

$0 + 18 =$ _____

$13 + 7 =$ _____

$11 + 2 =$ _____

Made in the USA
Columbia, SC
22 October 2024

44800555R00078